Editors-in-Chief

Prof. Janusz Kacprzyk
Systems Research Institute
Polish Academy of Sciences
ul. Newelska 6
01-447 Warsaw
Poland
E-mail: kacprzyk@ibspan.waw.pl

Dr. Lakhmi C. Jain
Adjunct Professor
University of Canberra
ACT 2601
Australia

And

University of South Australia
Adelaide
South Australia SA 5095
Australia
E-mail: Lakhmi.jain@unisa.edu.au

For further volumes:
http://www.springer.com/series/8578

Dennis Jarvis, Jacqueline Jarvis, Ralph Rönnquist, and Lakhmi C. Jain

Multiagent Systems and Applications

Volume 2: Development Using the GORITE BDI Framework

Authors

Dr. Dennis Jarvis
Centre for Intelligent and Networked
Systems
Central Queensland University
Rockhampton Queensland
Australia

Dr. Jacqueline Jarvis
Centre for Intelligent and Networked
Systems
Central Queensland University
Rockhampton Queensland
Australia

Dr. Ralph Rönnquist
Intendico Pty Ltd.
Carlton, Victoria
Australia

Dr. Lakhmi C. Jain
Adjunct Professor
University of Canberra
ACT 2601
Australia

ISSN 1868-4394   e-ISSN 1868-4408
ISBN 978-3-642-42858-6   ISBN 978-3-642-33320-0 (eBook)
DOI 10.1007/978-3-642-33320-0
Springer Heidelberg New York Dordrecht London

Foreword

In the history of Artificial Intelligence, the paradigms have their own lifecycle. They are born, mature and then die. Few of them are defying the time and through successive metamorphoses they transcend.

Such a paradigm was born in 1987, when the philosophy professor Michael Bratman published his first book *Intention, Plans, and Practical Reason*, introducing the belief-desire-intention (BDI) model for human practical reasoning. This served as an inspiration for the BDI software model for programming intelligent agents. The first system based on this model, the procedural reasoning system (PRS), was developed at SRI International by a team that included Michael Georgeff and Amy L. Lansky, and was still used at the end of the 90's. Many other systems followed with their strengths and weaknesses.

A special recent place amongst the systems inspired by the BDI paradigm is occupied by GORITE, which extends the paradigm to cover goal-oriented teams of intelligent agents. Inspired by long-life experience in multi-agent software design and implementation, GORITE is designed by professionals for professionals, as an extensible platform aiming to provide a flexible solution to the development of systems with complex interaction between their entities. Allowing the move of the control from the individual to the social, GORITE is also on the frontispiece of systems design and control research.

This book satisfies many needs of the professionals and educators in the multi-agent system. First, it is a very useful addition for any course in multi-agent systems that want to give the students a hands-on approach, through an integrative view, combining modeling strategies with implementation details. Second, it is a great introduction to GORITE for the professionals in multi-agent systems, allowing them a speedy immersion in GORITE which the immediate benefit of building complex applications with direct control and coordination over a team of agents.

I was always impressed by the long-life research performed in Artificial Intelligence by people who believed and developed innovative approaches and systems, over their career. Examples include ACT-R, started around 1973 by John R. Anderson and currently maintained by Carnegie Mellon University (act-r.psy.cmu.edu), SOAR, started around 1983 by Alan Newell, John Laird and Paul Rosenbloom and currently maintained at University of Michigan (sitemaker.umich.edu/soar/home), CYC, started around 1984 by Douglas Lenat and currently maintained by CYCORP (www.cyc.com), Disciple, started around 1986 by George Tecuci and currently maintained by Learning Agents Center of

George Mason University (lac.gmu.edu), and Protégé, started around 1989 by Mark
A. Musen and currently maintained by Stanford Center for Biomedical Informatics
Research at the Stanford University School of Medicine (protege.stanford.edu).

I wish to GORITE and its creators to enlist themselves on the same longevity
and accomplishments scale.

Mihai Boicu, Ph.D.
Associate Director of the Learning Agents Center
George Mason University
USA

Preface

Since its conception almost 30 years ago, the BDI (Belief Desire Intention) model of agency has become established, along with Soar, as the approach of choice for practitioners in the development of knowledge intensive agent applications. It has also been provided with a logical formalization, which has made it attractive to researchers. However, in developing BDI agent applications for over 15 years, we have observed a disconnect between what the BDI model provides and what is actually required of an agent model in order to build practical systems.

The focus of the BDI model is practical reasoning and it employs an intuitive combination of an individual agent's beliefs, desires and intentions for its achievement. This construction is generic, and can be applied to many types of agent reasoning, as evidenced by the diversity of implemented BDI applications. These applications have been supported by a progession of industrial strength software frameworks, beginning with PRS in the mid 1980's and continuing to the current day. In terms of support for BDI reasoning, these frameworks have traditionally provided an execution model in which goals, modelled as events, direct plan execution. Actual agent behaviour is specified within plans using a framework-specific plan language. Furthermore, since plans initiate goals, the specification of hierarchical agent behaviour becomes procedurally embedded in sub-plans.

While the above execution models have been extended in various ways – for example, JACK Teams provides support for the execution of team goals – the procedural embedding of hierarchical plan specification in sub-plans remains. Also, the ability for an agent to reason about intention is restricted to the choice of a course of action to achieve a goal or to the reconsideration of how a goal may be achieved if a course of action has failed. In practice, more sophisticated reasoning capabilities are required, particularly for those agents that are embedded in the real world and have a need to monitor the progress their own actions and those of others. The BDI execution model can be subverted to accommodate such tactical reasoning, but its underlying concept of reconsideration on plan failure presents a major stumbling block. As a consequence, such reasoning is usually delegated to plans and ad-hoc strategies implemented. Finally, the BDI model assumes that agents maintain their own beliefs, and that these beliefs are adjusted according to processes that are outside the scope of the model. From a framework perspective, this means that no distinction is made between beliefs that an individual agent holds and those that it shares when it is achieving a goal with other agents.

The GORITE framework was developed to address these perceived limitations, whilst still retaining the essence of the BDI model. In GORITE, this is achieved through an inversion of control – rather than an agent explicitly managing its own behaviour, the agent delegates that responsibility to an executor object, which then initiates goal executions on behalf of the agent. This change in perspective means that an agent's goals can be explicitly represented, directly manipulated and monitored. BDI execution semantics are preserved, with the agent still able to choose between courses of action to achieve a goal or to reconsider how a goal might be achieved. Team goals are specified in terms of roles which are filled by team members; team goal execution is then managed by a single executor on behalf of all the participating team members. Finally, the need to procedurally embed a plan's hierarchy in its sub-plans is removed, as the executor object traverses and executes explicitly defined goal/team goal hierarchies. During this process, the executor makes available to the participants in the execution a shared data context, thus providing for a clear separation between an agent's individual beliefs and those that it shares with other agents involved in the goal execution.

While GORITE remains true to its BDI heritage, it employs a quite different software architecture and execution model to its predecessors. While these aspects have been described in the research literature, we have written this book to provide a complete and coherent overview of the GORITE architecture, together with a comprehensive set of examples that explain how GORITE can be deployed in multi-agent application development. The real-world potential of multi-agent systems is emphasized with two of the three examples employed in the book focusing on sensor network management and manufacturing control. These examples highlight an observation that we have made from our work in the defence and manufacturing sectors – namely that the adoption of a team metaphor can bring a clarity to complex systems that tends to disappear if one adopts an individual agent perspective. As discussed in the book, it is interesting to see that this perspective, where a system is viewed as a collection of goals that are achieved by the explicit orchestration of teams of actors, is gaining credence in the wider software engineering community.

As GORITE is a Java framework, GORITE applications can be developed entirely in Java. Consequently, a familiarity with Java (or a similar language) is assumed, but no prior knowledge of the BDI model is required.

Acknowledgements

Prabin Rijal and Tony Martin contributed to the development of the examples presented in Chapters 3 and 5. Prabin's involvement was partially funded by the Queensland Government NIRAP Smart Future project "Solar Powered Nano Sensors", led by Nunzio Motta from the Queensland University of Technology. We would also like to thank the book's two reviewers – Mihai Boicu and Gheorge Tecuci, both from George Mason University – for their insightful comments and suggestions for improvement.

Contents

Acronyms

API	Application Program Interface
BDI	Belief Desire Intention
BPM	Business Process Modelling
CGF	Computer Generated Forces
DCA	Data Collaboration Algorithm
DCI	Data Context Interaction
FMS	Flexible Manufacturing System
HMS	Holonic Manufacturing Systems
IDE	Interactive Development Environment
JDK	Java Development Kit
MVC	Model View Controller
NPC	Non Playing Character
OODA	Observe Orient Decide Act
PLC	Programmable Logic Controller
RMI	Remote Method Invocation
SME	Subject Matter Expert
TOP	Team Oriented Programming
UAV	Unmanned Air Vehicle
UML	Unified Modelling Language
WIP	Work in Progress

Referenced Systems

America's Army	U.S. Army, 2012
BPMN	OMG, 2012
CoJACK	AOS Group, 2012
dMARS	d'Inverno et al., 1998
GORITE	Rönnquist, 2012
JACK	AOS Group, 2012
JACK Teams	AOS Group, 2012
JADE	JADE, 2012
OOram	Reenskaug, 1995
Prometheus	Padgham and Winikoff, 2004
PRS	Georgeff and Lansky, 1986
Soar	University of Michigan, 2012
TEAMCORE	Pynadath et al., 1999
VBS2	Bohemia Interactive, 2012

The Examples

Section	Example	Feature(s) illustrated
2.1	Hello World V1.1	Goal execution
2.1	Hello World V1.2	Data Context
2.1	Hello World V 1.3	Applicable set
2.1	Querying a Relation V1	Relation, Query
2.1	Hello World V1.4	Applicable set
3.2	Meter Box Cell V1.1	SequenceGoal
3.2	Meter Box Cell V1.2	LoopGoal
3.2	Meter Box Cell V1.3	RepeatGoal
3.3	Sensor Network V1.1	Baseline for single performer version
3.3	Sensor Network V1.2	FailGoal
3.3	Sensor Network V1.3	ParallelGoal
4.2	Meter Box Joining V1.1	Action goal, polling
4.2	Meter Box Joining V1.2	Action goal, blocking
4.3	Meter Box Cell V2	Perceptor, ToDo group manipulation
5.3	Hello World V2	Team, Role, supervised Task Team
5.4	Meter Box Cell V2	Team, Role, unsupervised Task Team
5.5	Sensor Network V2.1	Baseline team version
5.5	Sensor Network V2.2	Dynamic reformation of the task team
5.5	Sensor Network V2.3	Dynamic addition of team members
6.1	Querying a Relation V2	Query spanning two relations
6.2	Sensor Network V3	Monitoring a relation with a reflector
6.3	Fault Space Pruning V1	Belief propagation with new fact assertion
6.3	Fault Space Pruning V2	Belief propagation with fact retraction

Introduction

Wooldridge (2009) defines an agent as

> *"An agent is a computer system that is situated in some environment, and that is capable of autonomous action in this environment in order to meet its delegated objectives"*

As a definition, this is quite general, admitting systems such as thermostats and printer daemons into the world of agents. Consequently, Wooldrige then distinguishes between agents and intelligent agents, ascribing to intelligent agents the properties of reactivity, proactiveness and social ability. In this context, reactivity refers to an agent's ability to react to changes in its environment, whereas proactiveness refers to its ability to pursue an activity before it is required to be performed. Social ability lies at the heart of the multi-agent paradigm and relates to an agent's ability to achieve goals as part of a team.

The BDI (Belief Desire Intention) model has become, along with Soar, the method of choice in the development of intelligent agent applications (Jones and Wray, 2006). The key reasons for this are that the BDI model (as we shall see in Chapter 1) provides an intuitive basis for the modelling of agent behaviour and that the model has been supported by commercial strength software frameworks since its conception by Bratman in the 1980's.

Multi-agent systems have achieved commercial success in niche areas, most notably the augmentation of behaviour for entities in computer based war games. However, neither the BDI model nor the multi-agent paradigm in general has seen widespread application beyond these niche areas. It is our belief that the multi-agent paradigm is ideally suited to the development of complex systems where the complexity arises through the interaction between entities, rather than through the internal reasoning of individual entities. Examples of such systems are widespread and occur in domains as diverse as computer games, sensor networks, industrial control, UAV management and business process modeling. While we have reached this conclusion from our work in multi-agent systems, an interesting parallel is emerging in the software development community, with a resurgence of interest in roles (Coplien and Bjørnvig, 2010). As we shall see later, the concept of a role is central to the interaction model that underpins GORITE.

GORITE was developed by Ralph Rönnquist as the first step in a larger research program to explore the validity of our hypothesis. As such, it was designed from a multi-agent team perspective and it is implemented as a Java framework so that it is both accessible to developers and extensible. This book represents a second step in the program. Its intent is threefold:

1. to provide an accessible overview of the BDI model and its limitations from a software development perspective,
2. to explain how GORITE addresses those limitations and
3. to provide a practical introduction to the development of BDI applications using GORITE

In keeping with these aims, the book is structured as follows. It begins in Chapter 1 with an overview of multi-agent systems, the BDI model and an explanation of where (and why) GORITE is positioned in the BDI landscape. Chapter 2 introduces the reader to programming using GORITE through a series of variations on "hello world" with an alien flavour. Chapters 3 to 6 then explore the core aspects of GORITE – process models, situated action, teams and beliefs. These aspects are illustrated through examples drawn from the domains of manufacturing and sensor networks. Finally, Chapter 7 discusses future directions.

The GORITE framework is a work in progress – this book is based on the current release, which is v9RC04. The software is not open source, but is available to registered users under the terms and conditions of the Lesser GNU Public Licence – contact Ralph Rönnquist at ralph.ronnquist@intendico.com. All the examples in this book are available from the GORITE website – http://www.intendico.com/gorite.

Chapter 1
Multi-Agent Systems

Intelligent agent technology is at an intriguing stage in its development. Commercial strength agent applications are increasingly being developed in domains as diverse as manufacturing (Deen, 2003; Bussman et al., 2004; Jarvis et al., 2008a) war gaming (Jones et al., 1999; Heinze et al., 2002) and UAV mission management (Karim and Heinze, 2005). Furthermore industrial strength development environments are available, e.g. (AOS Group, 2012; University of Michigan, 2012; JADE, 2012) and design methodologies (Padgham and Winikoff, 2004) reference architectures (van Brussel et al., 1998) and standards (IEEE Computer Society, 2012) are beginning to appear. These are all strong indicators of a maturing technology. Furthermore, it has been proposed as the paradigm of choice for the development of complex distributed systems (Decker, 2004) and as the next step forward from object oriented programming (Wooldridge and Jennings, 1995). However, the uptake of the technology is not as rapid or as pervasive as its advocates anticipated.

In this chapter, we begin by looking at where multi-agent systems technology is positioned in terms of its commercial uptake. We will then provide an overview of the BDI model of agency which underpins GORITE and finally, we will introduce GORITE and the concept of team programming. It is our belief that team programming, with its twin focus on organizational structure and coordinated behaviour may offer a way forward for the multi-agent systems paradigm.

1.1 Two "Success" Stories

In this section, our discussion will focus on two areas – one, a niche area where multi-agents system have achieved commercial success (computer generated forces) and the other where extensive research activity over many years has not yet resulted in significant industry uptake of the technology (manufacturing). We will then consider what these two very different stories might hold for the future of the technology.

In the military domain computer generated forces (CGFs) are widely employed in both training and research. Systems such as VBS2 (Bohemia Interactive, 2012) provide a rich palette of entities (agents) at various organizational levels (e.g. soldiers, sections, platoons, companies) and a wide range of behaviours at varying degrees of autonomy (e.g. player controlled movement, inbuilt path-finding). Scenarios for training or research purposes can be constructed and played out

D. Jarvis et al.: Multiagent Systems and Applications, ISRL 46, pp. 1–12.
DOI: 10.1007/978-3-642-33320-0_1   © Springer-Verlag Berlin Heidelberg 2013

using built in, scripted or player provided behaviours as appropriate. Depending on the scenario, complex interplays involving autonomous groups of entities may be required. For example, a section may be required to retain formation while moving through difficult terrain or while they are under fire. Also, a company attack will require the coordination of fire platoon and attack platoons (Connell et al, 2003).

With VBS2 and its ilk, the norm is to provide access to the internal behaviour of these systems through mechanisms such as APIs, scripting languages or TCP/IP messaging. As a result, external processes are often able to control and augment the behaviour of CGF entities, in particular customizing entity behaviours to align with local military doctrine and terrain. The multi-agent paradigm has achieved considerable commercial success in this regard by using the CGF platform as essentially a visualization vehicle for individual entities. Entity behaviour, organizational structures and organizational behaviour are then implemented externally to the CGF platform in a separate multi-agent based application which interacts with the CGF platform through message passing (Lui et al., 2002; Connell et al., 2003; Jarvis et al, 2005).

The traditional use of computer generated forces has been at the theatre level – military resources move (or are moved) on a map of the area of engagement, as illustrated in Figure 1.1. However, given the continuing decrease in the cost/performance ratio in games technology, video games are becoming increasingly deployed in military training, as evidenced by the success of America's Army (U.S. Army, 2012). The success of this approach is critically dependent on the quality of the behaviour provided by the NPCs (non-playing characters) in the game. In conventional gaming, acceptable NPC behaviour has usually been able to be realized with simple techniques, such as state machines and flocking algorithms (Millington and Funge, 2009). However, military training imposes much more stringent realism requirements on NPC behaviour and multi-agent systems are being used to provide this realism. Recent examples of this include requirements for realistic group behaviour of NPC targets when fired upon (Jarvis et al., 2004), the incorporation of fatigue and fear into NPC behaviour (Evertsz et al. 2007) and the extension of military doctrine to incorporate legal and political considerations (Evertsz et al. 2009).

Likewise, the manufacturing domain has provided a similarly rich and varied playpen for the exploration of the multi-agent paradigm. However, while military multi-agent systems activity has focused on virtual applications such as simulation-based training, the focus for manufacturing has been the management of manufacturing activity. Despite significant industrial involvement in multi-agent systems research (e.g. the Holonic Manufacturing Systems (HMS) project (Brennan and Norrie, 2003)), the technology has not yet established a foothold in manufacturing. While there is a widely held belief that the key to realizing effective utilisation of resources in a dynamic, market-driven manufacturing environment lies in the adoption of the multi-agent paradigm, that belief has not been evidenced in practice. We suspect that the reason for this lies with the layered structure of manufacturing operations. Traditionally, manufacturing operations have been partitioned into three layers – planning, scheduling and

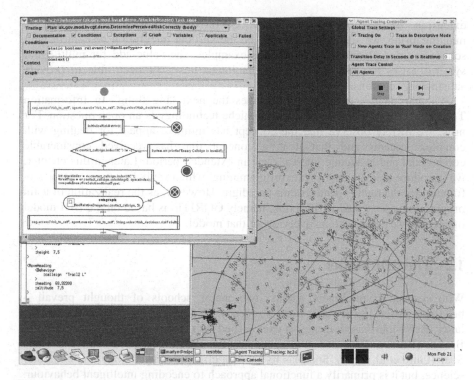

Fig. 1.1 A Snapshot of the MIG Vignette described in (Jarvis et al., 2005)

control. The problem is that these layers employ two quite different technologies and cultures. Planning and scheduling reside in the business world with applications running on conventional hardware, whereas control is essentially an engineering activity involving specialized hardware, such as PLCs.

While the technical feasibility of the multi-agent approach has been proven in each of the separate layers (Deen, 2003), the generic business case remains to be made. For example, the P2000+ agent-based control system was developed by DaimlerChrysler and it operated successfully for five years (Schild and Bussmann, 2007). However, DaimlerChrysler have not replicated the system. The reason given for this is that it was difficult to argue the business case. The major advantage of the P2000+ system was its flexibility, which is difficult to quantify and is often not required. The standard metric for manufacturing performance is throughput and while P2000+ provided improved throughput, other technologies could have done likewise for less cost. A similar experience was encountered in the Australian automotive industry (Jarvis et al., 2003).

The problem that faces agent-based manufacturing is that in each of the three layers there are existing technologies that are perfectly adequate for their purpose. Similarly in defence, systems like VBS2 are not being developed using agent-based technology – existing software technology is adequate. The challenge that faces the multi-agent systems community is to identify areas where the multi-agent paradigm provides a clear value add. In defence, the customization and extension of existing CGF frameworks provides one such example. In manufacturing, we suspect that it will be the vertical integration of control with scheduling and planning. In both these examples, the value-add for multi-agent systems arises because alternative technologies have not been able to provide satisfactory solutions.

Whether multi-agent systems becomes the next "big thing" in Information Technology or whether it will remain a niche technology is an open question. For now, we merely observe that the concept has intuitive appeal for dealing with complex systems and that it is a concept that has received considerable commercial and academic interest over an extended period. Later in this chapter, we introduce the concept of team programming, which we believe can offer a way forward for the multi-agent systems paradigm. However, as the realization of team programming presented in this book, namely GORITE, is based on the BDI model of agency, we begin with an overview of that model.

1.2 The BDI Model

With respect to intelligent agents, two major schools of thought prevail – behaviour-based agency and rational agency. We are strongly aligned with the latter. Behaviour-based agency is exemplified by the subsumption architecture of Brooks (Brooks, 1999); rational agency by the BDI model (Bratman, 1987) and Soar (Laird et al, 1987). Soar has its roots in cognitive psychology and computer science, but it is primarily a functional approach to encoding intelligent behaviour. The continuing thread in Soar research has been to find a minimal, but sufficient set of mechanisms for producing intelligent behaviour. Central to the Soar architecture are the problem space hypothesis (Newell, 1982) and the physical symbol systems hypothesis (Newell, 1980). Problem spaces modularize long-term knowledge so that it can be brought to bear in a goal-directed series of steps. The physical symbol systems hypothesis argues that any entity that exhibits intelligence can be viewed as the physical realization of a formal symbol processing system. Soar is characterized by the use of production systems for symbol processing and a uniform representation for knowledge and beliefs. While Soar imposes strong constraints on fundamental aspects of intelligence, it operates at a lower level than BDI frameworks in terms of reasoning.

The Belief-Desire-Intention (BDI) model is concerned with how an agent makes rational decisions about the actions that it performs. In the BDI model, agents have

- Beliefs about their environment, other agents and themselves.
- Desires that they wish to satisfy.
- Intentions to act towards the fulfilment of selected desires.

The model is characterized by having a philosophical basis, a rigorous logical formulation and a history of commercial strength implementations. These implementations include PRS (Georgeff & Lansky 1986), dMARS (d'Inverno, et al. 1998), JACK (Howden, et al. 2001; AOS Group, 2012) and most recently, GORITE (Rönnquist, 2012). As a consequence, the BDI model has underpinned many successful agent applications (Evertsz et al, 2004). Indeed, Jones and Wray (2006) have observed that BDI implementations have been one of the preferred vehicles (along with Soar) for the delivery of industrial strength, knowledge rich, intelligent agent applications.

Bratman's theory of human practical reasoning (Bratman, 1987), also known as the Belief-Desire-Intention theory, forms the philosophical basis for the BDI architecture for intelligent agents. The three attitudes of belief, desire and intention represent respectively the information, motivational and deliberational states of an agent (Rao and Georgeff, 1995). For Bratman, intentions are desires to which the agent has commitment. This commitment implies the temporal persistence of the intention and it leads to further plans being made on the basis of the commitment. Intentions must, in collection, be consistent and additionally should generally not be believed impossible. A goal becomes an intention when an agent adopts it – in BDI implementations this is taken to mean when the agent begins to act upon it.

In Bratman's theory, an agent divides its 'thinking' time between deliberating about its intentions, and planning how to achieve those intentions. This led him to identify three types of deliberation:

1. Goal deliberation, which is the process of generating a consistent set of goals, perhaps by selection from a set of desires.
2. Intention deliberation, which means choosing a goal (or goals) that the agent will act upon (and so will become an intention) and
3. Plan deliberation, which means constructing a plan, or selecting one from a plan library, that will further one or more of the agent's intentions.

Bratman further constrained deliberation by requiring the intentions to be consistent; this serves to limit the number of options considered and therefore to save time.

Another issue that is addressed in Bratman's theory is that of intention reconsideration, which means revisiting the deliberation process while the goal, intention or plan is still in effect. It can be beneficial to reconsider an intention – for example to exploit changes in the environment. There is a trade-off, however as reconsideration takes time. Insufficient reconsideration risks missing opportunities; reconsidering too often wastes time that could be used for planning or action. Thus Bratman also considers non-reconsideration. Under certain circumstances (typically when time is short) it is rational not to reconsider when confronted with changed conditions.

Bratman's philosophical model was first formalised by Rao and Georgeff (1991). They subsequently proposed an abstract architecture (Rao and Georgeff, 1995) in which beliefs, desires and intentions were explicitly represented as global data structures. Agent behaviour is event driven, and is realised through the following execution model:

```
initialize-state();
repeat
   options:= option-generator(event-queue);
   selected-options:= deliberate(options);
   update-intentions(selected-options);
   execute();
   get-new-external-events();
   drop-unsuccessful-attitudes();
   drop-impossible-attitudes();
end repeat
```

Rao and Georgeff observed that while their architecture was an idealization that faithfully captured Bratman's theory, it did not constitute a practical system for rational reasoning. In order to ensure computational tractability, they proposed the following representational changes:

- Only beliefs about the current state of the world are represented explicitly
- Information about the means of achieving certain future world states and the options available to an agent are represented as plans
- Intentions are represented implicitly by the collection of currently active plans

Rao and Georgeff did not discuss goal representation. In traditional BDI frameworks, goals are represented as events and have only a transient representation (Thangarajah et al, 2002), acting as triggers for plan invocations. Rao and Georgeff's execution loop then becomes

```
repeat
   wait for next goal event
   select (on the basis of current beliefs) a plan
      to achieve the current goal
   execute the selected plan
   update beliefs
end repeat
```

It is this loop that has provided the conceptual basis for all major BDI architectures (Pokahr et al., 2005). In this revised execution model (which we will refer to as the BDI execution model), plans consist of plan steps that are specified using a framework dependent plan language; these steps may involve the posting of further goal events. Note that more than one plan may be applicable for the achievement of a particular goal – this set of plans is called the applicable set. The selection of a plan to execute from the applicable set is based on the currently held beliefs of the agent and may involve explicit (i.e. meta-level) reasoning.

A key feature of Bratman's theory was the introduction of commitment as a characteristic that distinguishes intention from desire. In the context of BDI, the primary importance of commitment is in the temporal persistence that it gives to intention; intention selection is driven by desires (or, as in BDI implementations, goals). Traditionally, commitment is seen as a strategy that an agent employs in the reasoning processes associated with intention selection (Rao and Georgeff, 1991; Wooldridge, 2000). In particular, a commitment strategy is defined by Wooldridge as

"the mechanism that an agent uses to determine when and how to drop intentions".

Wooldridge identified the following strategies:

1. *Blind (or fanatical) commitment*: the intention is maintained until it is believed to have been achieved. For example: a captured seaport must be cleared of mines. Mine clearing proceeds until the port is safe for marine traffic. Setbacks such as mine explosions do not deter the main effort.

2. *Single-minded commitment*: the intention is maintained until it is believed either that it has been achieved or that it is impossible. For example: a platoon of soldiers attempts to occupy a strategic hilltop. The attempt is continued until it is successful or the platoon leader decides it is not possible for his platoon to achieve this objective.

3. *Open-minded commitment*: the intention is maintained until it is believed that it is impossible. For instance: a platoon maintains its position on a hilltop until the platoon leader decides it is no longer possible.

Traditional implementations of the BDI model do not concern themselves directly with commitment, preferring to focus on beliefs, desires and intentions as the primary mental attitudes of interest. Wooldridge demonstrates how the commitment strategies listed above could be incorporated into BDI implementations, but his approach would result in an agent's behaviour being characterized by a particular strategy. While such an approach has yielded interesting and useful insights into agent behaviour (e.g. (Rao and Georgeff, 1991)), in practice, it is insufficient to capture the subtleties of agent behaviour, particularly when agents are required to exhibit realistic human-like behaviour. For example, an intention might be dropped because it is found to be inconsistent with other (more important) intentions or because an overarching intention has been reconsidered and dropped. It may also be the case that intentions are suspended while other more urgent intentions are pursued. Neither of these commonplace reasons for the dropping of intentions is considered by Wooldridge in his characterization of commitment. To avoid confusion, we use the term commitment strategy to refer to one of the three strategies identified by Wooldridge and the term intention management to refer to the processes that an agent or group of agents uses to determine when and how to adopt and drop intentions.

In practical BDI systems, intention management becomes a major behaviour modelling issue. For example, in war-gaming applications (e.g. Connell et al., 2003; Jarvis et al., 2005), the progression of a platoon towards a waypoint that must be reached by a particular time may be impacted on by the observation of enemy troops. Does the platoon continue directly towards the waypoint? Do they hide? Do they take an alternate route? Or do they abort and return to base? Each option represents different levels of commitment to the current intention. Similar issues arise in engineered systems. For example, in a manufacturing cell, tool breakage needs to be detected and contingency plans activated. It may be that the tool can be replaced or the particular operation can be performed by another machine. Alternatively, the current job could be aborted and another job (not requiring the broken tool) selected or perhaps all processing could be aborted until the tool is replaced. Likewise, if resource contention is an issue in a manufacturing cell, reasoning about what operation to perform next may need to be made on a step by step basis. Such practicalities are important if an agent is to exhibit realistic behaviour. However, the BDI model itself provides no guidance or support as to their realization. It is left to developers to implement appropriate intention management strategies using the programming constructs provided by the BDI framework being employed. Furthermore, the developer is required to implement these strategies in environments where intention is not explicitly represented.

We would argue that intention management is an integral component in the specification of BDI behaviours. That is, essential concerns for a BDI agent are when to continue or discontinue with the current intention and how the current intention is resourced. The implication of this perspective is that the progress of intentions needs to be monitored against appropriate maintenance conditions and that intentions need to be explicitly manipulable (i.e. deleted, added, prioritised, suspended and resumed). The BDI execution model does not support the monitoring of intention (viewed either as goal or plan instances). Also, plans that provide meta-level reasoning can be triggered, but such reasoning is restricted to plans in the applicable set, not the plans that the agent is currently pursuing or are scheduled to be pursued.

In summary, the BDI execution model supports neither the monitoring of intention nor the explicit manipulation of intention, both critical aspects of intention management. Furthermore, intention management needs to operate at both an individual agent level and at a team level. In the next section, we discuss the BDI model from an agent team perspective.

1.3 Team Programming

In artificial intelligence, philosophy and cognitive science, there is general consensus that the collective activity of a group is more than the aggregate of the domain-oriented actions performed by the individuals belonging to the group. However, there is ongoing debate as to how collective behaviour should be modelled. A key issue is whether collective intentionality requires a different kind of mental-state construct, namely an intentional attitude that although individually

held is different from and not reducible to an "ordinary" intention. Opposing
views have been presented by (Searle, 1995) (for) and by (Bratman, 1999)
(against). From a multi-agent perspective, this tension is reflected in the Joint
Intention theory of Cohen and Levesque (Cohen and Levesque, 1991) and the
SharedPlans theory of Grosz and her collaborators (Grosz and Kraus, 1996).

   In the Joint Intention theory, a team is defined as "*a set of agents having a
shared objective and a shared mental state*". Joint intentions are held by the team
as a whole, and require each agent to commit to informing other team members
whenever it detects that the common goal has been achieved, or has become
impossible to achieve, or that because of changing circumstances, the goal is no
longer relevant. By way of contrast, individuals in the SharedPlans theory deploy
two intentional attitudes – *intending to* do an action and *intending that* a
proposition holds. The intending to attitude is used to represent an agent's
commitment to its own actions; the intending that attitude is used for group
activity and the actions of its team members in service of that activity. Such
commitments lead an agent to engage in what Grosz refers to as intention
cultivation. In this activity, an agent reasons about the actions and intentions of its
fellow team and it determines ways that it can contribute to their success in the
context of group activity. Both Joint Intention theory and SharedPlans theory have
provided the theoretical basis for successful implementations of team behaviour,
most notably the team oriented programming (TOP) model of TEAMCORE
(Pyandath et al., 1999). The TOP model is based on Joint Intentions theory and
incorporates aspects of SharedPlans.

   Other implementations not grounded in the above theories include cooperation
domains and JACK Teams. The concept of cooperation domains, as implemented
in (Tamura et al, 2003), encapsulates the interaction that occurs over the lifecycle
of a team created to execute a particular task. Team member behaviours are
external to the cooperation domain; the cooperation domain only provides the
protocol for team interaction. While cooperation domains adopt a similar
modelling viewpoint to that of the SharedPlans and Joint Intentions theory where
team behaviour is an individual agent concern, in JACK Teams (AOS Group,
2012) team behaviour is modelled separately from individual behaviour, using
explicit team entities. The adoption of this approach has provided significant
benefits from a software engineering perspective by enabling team behaviour to be
modelled and understood as a top-down process, rather than a bottom-up process
as in other approaches.

   JACK Teams was the first framework to extend the BDI model of agency to
include teams as explicitly modelled entities – that is, a team has its own beliefs,
desires and intentions and is able to reason about team behaviour. Team behaviour
is specified independently of individual behaviour through the use of roles. A role
in JACK Teams is an interface definition that declares what an entity that claims
to implement a role must be capable of doing. It has two aspects – the goals
involved in fulfilling the role and the beliefs that are to be exchanged. A team
member commits to the performing of a particular role – there is no notion of
shared intention from the team member's perspective or of commitment to other
team members.

Prior to JACK Teams, team behaviour was viewed as an individual agent concern – that is, an agent possesses the capability of either acting individually or participating as a member of a team. Consequently there is no notion of a team as a distinct software entity that is able to reason about and coordinate team member behaviour. For example in TEAMCORE, joint intentions are represented as plans, but each individual agent executes its own copy. Reasoning regarding team behaviour is the responsibility of the individual team members and infrastructure is provided to propagate changes in mutually held beliefs between team members. A focus on interaction at the expense of behaviour is also evident in the concept of cooperation domains (Tamura et al, 2003) and the work of FIPA (IEEE, 2012).

Treating team behaviour as an individual agent responsibility has intuitive appeal, as humans do not separate the teaming aspects of reasoning from the individual aspects. For example, if a human is capable of performing a welding role, then he or she might execute reasoning along the following lines:

"If I am member of the VQ Shuttle production team, then if I am working with a partner, then I will be responsible for the welds on one side of the vehicle. However, if I am working alone, I will be responsible for both sides."

and

"If I am a member of the rework team, then I am responsible all welds on the vehicle".

However, such an approach introduces significant difficulties from a software engineering perspective, as the development of behaviours becomes extremely complex and brittle. For example, if an individual agent can assume many roles within a variety of teams, changes to the potential team structures will result in extensive redesign and the agent code quickly becomes unmanageable. JACK Teams overcame these difficulties by declaring teaming aspects separately from agent aspects. The role concept then provides the means to connect the two aspects. This declarative specification of team behaviour in a single modelling construct (the team plan) reflects a heritage of military tactics modelling (Lui et al, 2002; Heinze et al, 2002), where a critical issue is Subject Matter Expert (SME) involvement in tactics capture and validation.

The philosophical implications of representing a team as a distinct entity are only beginning to be explored. In (Jarvis et al, 2007), JACK Teams is positioned within the context of the conventional approaches to team behaviour and its relationship with Koestler's notion of a holon (Koestler, 1967) is examined. Koestler did not distinguish between a team and an individual – rather, he described system behaviour in terms of holons. As defined by Koestler, a holon is an identifiable part of a system that has a unique identity yet is made up of subordinate parts and in turn is part of a larger whole. Holons can then be thought of as self-contained wholes looking towards the subordinate level and/or as dependent parts looking upward. This duality can be equated at a software level to the JACK Teams role concept and at a philosophical level to Grosz's intentional attitudes of *intending-to* and *intending-that*.

While GORITE shares with JACK Teams the representation of teams as explicitly modelled entities with their own beliefs, desires and intentions, it does this from a very different perspective. JACK Teams extends the BDI model to accommodate team behaviour, but does so from a traditional BDI perspective. That is, it is grounded in the BDI execution model presented in the previous section and the representational framework of that model. In particular, goals are not represented explicitly and behaviour is specified procedurally using a plan language. In GORITE, a different representational framework is employed – one where goals are represented explicitly and behaviour is represented in terms of goal-based process models.

The initial motivation for the development of GORITE was to provide a goal-based modelling framework for business process modelling (BPM). From a BPM perspective, an organisation can be viewed as a single business recursively divided into business units, each of which operate with respect to common business objectives[1]. Structurally, a team/sub-team decomposition is a good match and a team-based BDI framework provides a sensible starting point. However, what distinguishes BPM from other team-based BDI applications is the coherence of the overall behaviour – entities may exhibit autonomy, but that autonomy is constrained to conform to organisational procedures and processes. Conventional BDI frameworks can of course be employed to realise such behaviour, but the supported modelling stance is one of unconstrained autonomy.

GORITE supports the coherence required of BPM through its notion of explicit goal based process models that, as we shall see in Chapter 5, are able to seamlessly integrate organizational structure with organizational behaviour. However, equally importantly the process models are executed with respect to a data context which is available to and modifiable by all entities involved in the process model realization. As such, GORITE supports a team programming metaphor in which a system is viewed as a collection of goals that are achieved by teams of agents that work together on a common data context according to well-defined business processes. By having a common data context (as in the order object that we use in our manufacturing examples later in the book) in addition to explicit representations of both intention and orchestration, a process model, unlike plans in traditional BDI frameworks, becomes a standalone entity in terms of the encapsulation of team behaviour. Thus the design focus shifts from providing each agent with the ability to participate in group behaviours to defining the group behaviours that are required for a particular application.

We believe that the team programming metaphor has application beyond business process modelling, but acknowledge that the benefits of the approach are best showcased by applications that exhibit complex behaviour through a need for intelligent coordinated action (as in manufacturing operations (Bussmann et al,

---

[1] Alternatively, an organisation may be viewed as a collection of separate entities that pursue separate business objectives whilst providing services for each other. GORITE provides an outsourcing model to support such a modelling perspective (Rönnquist and Jarvis, 2009) – the model is discussed in Chapter 7.

2004) or military operations (Jarvis et al., 2005)). Other application domains that could benefit from the approach include sensor networks, business supply chains and UAV mission management. Note that while we have arrived at the above position from a starting point of intelligent agent technology and in particular BDI frameworks, a similar synthesis is emerging in the world of object-oriented programming, led by Trygve Reenskaug, the originator of the MVC (Model View Controller) paradigm. Reenskaug was an early advocate for role-based modelling, with OOram (Reenskaug, 1995) and this work had a strong influence on the formulation of collaborations as they are presented in UML. Reenskaug is now advocating that complex system behaviour is best specified in terms of collaborations and roles and has developed the DCA (Data-Collaboration-Algorithm) paradigm to support this thesis (Reenskaug, 2007). DCA has subsequently evolved into the DCI (Data-Context-Interaction) paradigm that is being championed by Coplien and Bjørnvig (2010). Further discussion of DCI and its relationship to GORITE is deferred until Chapter 7.

Chapter 2
Getting Started with GORITE

Goal Oriented Teams (GORITE) is a Java framework for the implementation of goal oriented process models in a team oriented paradigm (Rönnquist, 2012). Both the design and implementation of GORITE is representative of industry best practice. The framework is also the subject of ongoing maintenance and enhancement – the current release (June 2012) is v9RC04. Given its primary design objective, GORITE is provided as a Java class library and there is currently no graphical development environment. However, in contrast to conventional BDI frameworks, there is no separate plan language – all development is in Java. Consequently, developers can continue to use their favourite IDE.

From a programming perspective, the GORITE API conforms, where practical, to JDK1.4. The reason for this is to make porting to new platforms more straightforward. When developing applications, developers are of course free to use features from the current version of Java. The GORITE API is documented using JavaDoc, and excerpts are available in this book as Appendix B. There is no separate user documentation – this book represents an attempt to fill that void.

With the exception of the examples presented in this chapter, the examples in this book are incomplete in that package statements, import statements and constant definitions are omitted. However, working code for all the examples is available from the GORITE website (http://www.intendico.com/GORITE). With respect to constants, we use the convention that their names are specified in upper-case. Exception handling is included when necessary, either by throwing or catching the exception. However, in the latter case, no behaviour is specified for when the exception is caught. Also note that output is sent to System.err, not System.out. The reason for this is that System.err is unbuffered, so output appears immediately it is initiated.

2.1 Hello World

"Hello World" represents, since its introduction in (Kernighan and Ritchie, 1978) the archetype for a first program when presenting a new programming language. We will not break with tradition here, but what we will do is introduce, in addition to a canonical version, a number of variants. The purpose of these variants is to illustrate, at an introductory level, how the GORITE execution model operates and how the key BDI concepts of beliefs, desires and intentions are represented. The rationale for this is that the representational structure and the execution model together provide a convenient mechanism for characterizing BDI frameworks.

D. Jarvis et al.: Multiagent Systems and Applications, ISRL 46, pp. 13–29.
DOI: 10.1007/978-3-642-33320-0_2   © Springer-Verlag Berlin Heidelberg 2013

2.1.1 Version 1: Goal Execution

In this example we will create an alien performer (a.k.a. agent) that can achieve a
Greetings goal by printing the message "Hello World". The initial application
consists of three source files – Alien.java, Greetings.java and Main.
java. The code for the Alien class is shown below:

Alien.java:

```
package ch2.greetings.v1a;
import com.intendico.gorite.*;


public class Alien extends Performer {

  // constructor
  Alien(String name) {
    super(name);
  }

  // A method to initiate a greeting goal
  public boolean greet() {
    // Greet the people
    return performGoal( new Greetings(),"Greetings",new Data() );
  }
}
```

This code resides in a file called Alien.java. The alien will be created from a
main() method contained in Main.java. The alien's greet() method will
also be invoked from main(). The Greetings goal is specified in
Greetings.java. Listings for Main.java and Greetings.java are
presented below.

Note that in the code above, the Alien class extends com.intendico.
gorite.Performer. The Performer class has a constructor which has a
single parameter – the name of the performer. It also makes available the
performGoal() method that is employed within the subclass for goal execution.
The first parameter for performGoal() is the goal instance to be executed. The
second parameter is used in execution tracing[1] – it provides the "head" for the
goal/process model that is being executed. In this case, the process model is a single
goal. The final parameter holds the data context for the goal execution. If the goal
execution requires access to any data elements for its completion, these elements
would be added to the data context. In this example, no application data is required
and therefore no data elements are added to the data context.

Goal execution is not performed directly by the performer. Rather, performers
delegate execution to one or more instances of the Executor class. An executor
traverses a process model, executing sub-goals as dictated by the model structure.

---

[1] Tracing is discussed in Appendix A.

In this example, the default executor is used – when required, executors can be explicitly created and bound to performers. Among other things, the executor makes the data context available to each sub-goal when it is executed. Execution of the Greetings goal is achieved by the invocation of its execute() method.

The code for the Greetings class is as follows:

Greetings.java:

```
package ch2.greetings.v1a;
import com.intendico.gorite.*;

public class Greetings extends Goal {

  // name of the goal
  public static final String GREETINGS = "greetings";

  public Greetings() {
    super( GREETINGS );
  }

  public States execute( Data d ) {
    System.err.println( "Hello World" );
    return States.PASSED;
  }
}
```

The Greetings class extends com.intendico.gorite.Goal and overrides its execute() method. As noted above, no data context is required for this example, so it is not accessed in the body of the execute() method. execute() returns a value indicating the state of the execution. In this case, execution has successfully completed, so the value PASSED (specified in the com.intendico.gorite.Goal.States enumeration) is returned to the performer's executor.

The file Main.java contains the application's main() method, which is responsible for creating an Alien object and invoking its greet() method:

Main.java:

```
package ch2.greetings.v1a;

public class Main {
  public static void main(String [] args) {
    Alien paul = new Alien( "Paul" );
    paul.greet();
  }
}
```

To compile and run the program, ensure that `gorite.jar` is included in the `CLASSPATH` variable. Then from the command line, you could enter the following lines:

```
javac Main.java Alien.java Greetings.java
```

and

```
java gorite.examples.alien.alien1.Main
```

Alternatively, you could employ the services of your favourite IDE.
    Either way, the following output is produced:

```
Hello World
```

Note that in the above example, the `Greetings` goal is specific to the alien performer and it requires only one instance of the goal. This situation is quite common in GORITE applications, particularly when team-based goals/process models are involved. It also means that anonymous classes[2] can be employed to reduce the number of files required by an application. For example, the above application can be collapsed to a single file (`Main.java`) through the use of anonymous classes and non-public classes, as shown below.

Main.java:

```
package ch2.greetings.v1b;

import com.intendico.gorite.*;

public class Main {
  public static void main(String [] args) {
    Alien paul = new Alien( "Paul" );
    paul.greet();
```

---

[2] An anonymous class is a class that is declared locally within a block of Java code and does not have a name. Since anonymous class definitions are expressions, they can be used for the definition of one-shot classes at the point that they are required. The syntax is as follows:

```
new classname(argument-list) { class-body }
```

If *class-name* is the name of a class, the anonymous class is derived from the named class. If it is an interface then the anonymous class implements that interface and is derived from `Object`. Since the anonymous class itself does not have a name, it is impossible to define a constructor and instance initialisers are used instead for object initialization. An instance initialiser is a block of Java code delimited by curly brackets.

```
    }
  }

  // there can be only one public class per Java file
  class Alien extends Performer {
    // constructor
    public Alien( String name ) {
      super( name );
    }

    // A method to initiate a greeting goal
    public boolean greet() {
      // Greet the people
      return performGoal(
        new Goal( "greetings" ) {
          public States execute(Data d) {
            System.err.println( "Hello World" );
            return Goal.States.PASSED;
          }
        },
        "Greetings",
        new Data()
      );
    }
  }
}
```

Whether one chooses to adopt this style is an individual design decision. However, it is a style that is used extensively in the examples provided with the GORITE distribution because of its compactness.

2.1.2   Version 2: Data Context

In this example, Paul tailors his greeting to the particular planet that is being visited. Thus, rather than saying "Hello World", if he was visiting earth, he would say "Hello people of Earth". In order to do this, Paul needs to know which planet is being visited. This information is made available to his greetings goal via the data context for the goal execution. The data context is of type com. intendico.gorite.Data. A Data instance is a container for data elements (of type Data.Element). A data element is essentially a name/value pair, but as we shall see in later examples, an element can assume multiple values. The reason for making data elements multi-valued is to provide support for and in fact enforce temporal awareness within a performer's reasoning processes. Thus, rather than forgetting the prior value when a new value is assigned, a performer automatically remembers all value assignments, in the order that they have been made. A performer may then use this history to guide its onwards reasoning processes.

As noted in the previous example, the performer's executor makes the data context available to each sub-goal in a goal execution. A given sub-goal can add and modify data element values, as well as access existing values. In this example, the data element with a name of "planet" is introduced into the data context and its value is then accessible within the `execute()` method of Paul's greetings goal. The listings for the modified example 1 files are presented below, with the modifications shaded.

Alien.java:

```java
import com.intendico.gorite.*;

public class Alien extends Performer {

  // constructor
  Alien(String name) {
    super( name );
  }

  // A method to initiate a greeting goal
  public boolean greet(String planet) {
    // Greet the people
    Data data = new Data();
    data.setValue( "planet", planet);
    return performGoal( new Greetings(),
      "Greetings", data );
  }
}
```

Greetings.java:

```java
package ch2.greetings.v2;
import com.intendico.gorite.*;

public class Greetings extends Goal {

// name of the goal
  public static final String GREETINGS = "greetings";

public Greetings() {
  super( GREETINGS );
}

public Goal.States execute( Data d ) {
  String p = (String) d.getValue( "planet" );
  System.err.println( "Hello people of "+p );
  return Goal.States.PASSED;
}
}
```

Main.java:

```
package ch2.greetings.v2;

public class Main {
  public static void main(String [] args) {
    Alien paul = new Alien( "Paul" );
    paul.greet( "earth" );
  }
}
```

The application, when run, produces the following output:

```
Hello people of earth
```

2.1.3 Version 3: The Applicable Set

As indicated in Chapter 1, a BDI framework can be characterised by its

1. representational framework, that is how beliefs, desires and intentions are represented and its
2. execution model.

The two previous examples have introduced two key aspects of the GORITE execution model, namely that

1. a performer delegates execution to a separate execution object and that
2. the executor can make available a data context for a particular goal execution which can then be accessed and modified by any of the sub-goals involved in the goal execution.

In terms of representation, goal execution in GORITE operates on explicitly represented goal (desire/intention) models. However, in the examples thus far, the goal model has consisted of a single goal (Greetings) that is achieved by a single performer (Paul). In the remaining examples in this chapter, this simplification is retained; goal hierarchies and teams of performers are introduced in later chapters. In the next two examples, we introduce BDI execution semantics. As before, these concepts are introduced within the context of the single performer Hello World example, so the treatment is by necessity introductory.

The essence of BDI execution lies in the concept of the *applicable set*. When a BDI agent elects to pursue a goal in the traditional model it considers all plans that could be used to achieve the goal that is to be pursued. This collection of plans is called the applicable set. In GORITE, there is no conceptual distinction between goals and plans – a plan is a goal that, as we shall see in the next example, can have context and precedence. The applicable set therefore contains goals and any goal that the agent can perform that has the same name as the goal that is to be achieved is included. Goals in the applicable set are attempted in sequence; execution stops as soon as a goal succeeds or when all goals have failed. When a

goal fails, the applicable set is regenerated before execution continues. The reason
for this is that if the applicable set contains plans, then the context for execution of
these plans may have changed while the failed goal was being pursued.

In our previous examples, an individual goal instance was created at the point
of execution, that is when performGoal() was invoked. However, for BDI
execution multiple goal instances (the applicable set) need to be considered and
those instances need to be persistent within the application. In GORITE,
persistence is achieved through the notion of a *capability,* which is a container of
goals (and as we shall see in Chapter 6, rules). A goal is accessible by its name;
multiple goals may be returned for a given name. These goals are understood to
represent alternative ways in which the named goal might be achieved.

The Performer class extends the Capability class, so all performers have
a default capability. Capabilities can also be created independently of performers
and added to performers. As you would expect, capabilities can form hierarchies.
Explicit capabilities will be considered in Chapter 5; until then, our examples will
utilize the performer's default capability.

Goal instances are added to the default capability using the addGoal()
method, which is normally invoked from the performer's constructor. For
example, the greetings goal of our first two examples could be added to the default
capability of the performer by including the following statement in the
performer's constructor:

```
addGoal( new Goal( "greetings" ) {
  public States execute(Data d) {
    System.out.println( "Hello World" );
    return States.PASSED;
  }
});
```

To execute a goal defined in a capability, we utilize a BDI goal as shown below. A
BDI goal looks up the given goal name in the performer's capabilities and
constructs an applicable set, which in the example above, will consist of a single
greetings goal. BDI goal execution then results in the members of the applicable
set being attempted in sequence with execution stopping as soon as a goal
succeeds.

The greetings goal added to the default capability above can then be executed
in the following manner:

```
return performGoal(
  new BDIGoal( "greetings" ), "Greetings", data );
```

Alternatively, BDI goal creation can be left to the `performGoal()` method:

```
return performGoal( "greetings", "Greetings", data );
```

This statement has the same effect as the previous statement.

In our next example, which will illustrate BDI execution semantics, Paul does not use the planet data provided in the data context to construct his message. Rather, he has two predefined greetings goals – one for greeting the people of Earth and one for greeting the people of Mars. The two goals both have the same name, so they will both be added to the applicable set. They are also (by default) of the same precedence, so they will be tried in the order that they are added to the performer. The success or failure of the goal execution is determined by comparing the value of the planet data element provided in the data context with the name of the planet for which the goal is configured.

Main.java:

```
package chapter2.greetings.v3;

public class Main {
  public static void main(String [] args) {
    Alien paul = new Alien( "Paul" );
    paul.greet( "mars" );
  }
}
```

Alien.java:

```
package chapter2.greetings.v3;
import com.intendico.gorite.*;

public class Alien extends Performer {

  // constructor
  Alien( String name ) {
    super( name );
    // Add goals
    addGoal( new GreetEarth() );
    addGoal( new GreetMars() );
  }

  // A method to initiate a greeting goal
  public boolean greet( String planet ) {
    // Greet the people
    Data data = new Data();
    data.setValue( "planet", planet );
    return performGoal(
      new BDIGoal( Greetings.GREETINGS),
        "Greetings", data );

  }
}
```

Greetings.java:

```java
package chapter2.greetings.v3;
import com.intendico.gorite.*;

public class Greetings extends Goal {
  // name of the goal
  public static final String GREETINGS = "greetings";

  public Greetings() {
    super( GREETINGS );
  }
}

class GreetEarth extends Greetings {

  public GreetEarth() {
    super();
  }

  public States execute(Data d) {
    String p = (String) d.getValue( "planet" );
    if ( p.equals( "earth" ) ) {
        System.err.println( "GreetEarth: Hello
          people of Earth" );
      return States.PASSED;
    }
    System.err.println( "GreetEarth: goal failed" );
    return Goal.States.FAILED;
  }
}

class GreetMars extends Greetings {

  public GreetMars() {
    super();
  }

  public States execute(Data d) {
    String p = (String) d.getValue( "planet" );
    if ( p.equals( "mars" ) ) {
        System.err.println( "GreetMars: Hello people
          of Mars" );
        return Goal.States.PASSED;
    }
    System.err.println( "GreetMars: goal failed" );
    return States.FAILED;
  }
}
```

When run, the program produces the following output:

```
GreetEarth: goal failed
GreetMars: Hello people of Mars
```

2.1.4 Version 4: Beliefs and Plans

In the previous example, Paul did not make use of the planet information at his disposal to customize a single greetings goal, as he did in the second example. Rather, he chose to pursue all (two) applicable goals in a fixed sequence. If the first goal passed, all was fine. If it failed, he tried the second goal. This approach assumes that the data that informs the decision (in this case, the name of the planet) is fixed and is provided in the data context for goal execution.

The data context is intended to model data which is specific to a particular goal execution. For example, in a manufacturing scenario, a customer order ("make 100 red widgets by Friday") could sensibly be modelled in terms of a data context element. However, it is unlikely that the properties of particular machines (e.g. status or capability) involved in fulfilling the order would be modelled in this way. Rather, it is likely that each machine would maintain its own persistent[3] data relating to its beliefs about itself and its environment.

Agent beliefs are modelled in GORITE as relations; a relation is a set of tuples. A relation may have one or more *key constraints*. A key constraint combines one or more fields into a *key*, which must be unique for each tuple in a relation. GORITE also supports the use of logic programming for belief reflection. This part of the GORITE framework includes the Ref class to represent logical variables, and a Query interface representing the abstract predicate.

The power of the BDI approach in terms of modelling behaviour lies in its ability to enable an agent to select a course of action to achieve its current goal on the basis of its beliefs about the world in which it is situated. In traditional BDI frameworks courses of action are specified in terms of constructs called *plans* which are triggered in response to goals represented as transient events. In GORITE, this representational disconnect of goals and plans does not exist[4] – plans are goals that have both context and precedence. Context in this sense refers to plan context (as opposed to data context) and refers to the use of a logical predicate (represented in GORITE as a Query) to determine both plan applicability and to dynamically generate plans corresponding to each logical binding of the context predicate. These plans are added to the applicable set and the binding for a particular plan is made available through the data context for the goal execution.

---

[3] By persistent, we mean that the data is relevant for longer than a single goal execution.

[4] From a BDI perspective, intentions span both goals and plans – an intention represents a commitment to achieve a particular goal. Plans are then chosen to achieve a goal. In GORITE, this relationship is captured by modelling plans as goals, but with precedence and context. In traditional BDI frameworks, the two concepts are representationally distinct, with goals represented as transient events which trigger plan execution.

In our final example in this chapter, we model Paul's belief as to his whereabouts as a tuple with a single field – the planet on which he has landed. The context method for the `Greetings` plan will then query the relation and add the binding to the data context, which is then accessible to the goal's `execute()` method.

For clarity, we will present this example in two parts. In Part 1, we illustrate how simple query processing in GORITE works through the use of a simple example. In Part 2, the standalone query presented in Part 1 is incorporated into the context method for the `Greetings` plan.

Part 1: Beliefs

In this example, Paul maintains beliefs relating to greetings that can be delivered to various planets. These messages are stored in a relation called `greetings` and its tuples are of the form

```
<planet, language, greeting>
```

The relation is created and populated in the constructor for the `Alien` class and only a simple query is provided – the `greeting()` method constructs a query to find all greetings relevant to a specified planet. More sophisticated query processing is supported – for example, an observer can be bound to a query and queries can be formulated as complex logical expressions. These topics are discussed in Chapter 6.

Alien.java:

```java
package chapter2.greetings.v4a;
import com.intendico.gorite.*;
import com.intendico.data.*;

public class Alien extends Performer {

  Relation can_greet
  // allow c[] to be declared as a raw (unparameterised) type
  @SuppressWarnings( "rawtypes" )

  Alien( String name ) throws Exception {
    super( name );
    // Beliefs about languages spoken
    // tuples are of the form <planet,language,greeting>
    Class c[] = { String.class, String.class, String.class };
    can_greet = new Relation( "can greet", c );

    // Set greetings
    can_greet.add( "mars", "martian", "VDREW^% ^%FD$^");
    can_greet.add( "earth", "english", "hello world");
    can_greet.add( "earth", "french", "bonjour le monde");
  }

  public Query greeting( String planet ) throws  Exception {
```

```
    // find all tuples that relate to the specified planet
    return can_greet.get( new Object [] {
      planet, new Ref("$language"), new Ref("$greeting") } );

  }
}
```

The actual binding of values for a query is separate from the query construction in the greeting() method above. Typically, binding will involve the following steps:

1. Provide the query instance with a collection of objects of type Ref to hold successive bindings. The query's getRefs() method is used for this purpose.
2. If necessary, revoke any previous query processing via the query's reset() method
3. Generate a binding by invoking the query's next() method. Note that a query may have multiple bindings; these are generated by additional calls to next(). When all bindings are exhausted, next() returns false.

This is demonstrated in the code below.

Main.java:

```
package chapter2.greetings.v4a;
import java.util.*;
import com.intendico.data.*;

public class Main {
  public static void main( String [] args ) throws Exception {
    Alien paul = new Alien( "Paul" );

    // Create the query. No binding has occurred.
    Query q = paul.greeting( "earth" );
    System.err.println( "Query = " + q );

    // The getRefs() method provides the query with the
    // collection of Ref objects that will be used to hold
    // (successive) query bindings.
    Vector<Ref> refs = q.getRefs( new Vector<Ref>() );

    // Review the Ref objects of Query q
    review( q, refs );

    // Review the Ref objects again - they will be null
    review( q, refs );

    // And again, but after a reset. The reset() method resets
    // the query to its initial state - the next invocation of
    // the next() method will provide the first set of bindings
    // for the query. It is not required to generate the first
    // set of bindings but it is necessary to regenerate the
    // bindings.
```

```
            q.reset();
            review( q, refs );
    }

    static void review( Query q, Vector  refs) throws Exception {
        // The first invocation of next()establishes the first valid
        // combination of bindings for the Ref objects involved, and
        // subsequent calls establish subsequent valid bindings. The
        // method returns false when it exhausts the valid
        // combinations of bindings.
        System.err.println( "Bindings are:" );
        int b = 0;
        while ( q.next() ) {
            System.err.println(" binding "+b++ );
        for ( Iterator<Ref> i = refs.iterator(); i.hasNext(); ) {
            Ref r = i.next();
            System.err.println( "    Ref "+r.getName()+" = "+r );
        }
        }
    }
}
```

When the program is run, the following output is produced:

```
Query = greetings( earth, $language=null, $greeting=null )
Bindings are:
 binding 0
    Ref $language = english
    Ref $greeting = hello world
 binding 1
    Ref $language = french
    Ref $greeting = bonjour le monde
Bindings are:
Bindings are:
 binding 0
    Ref $language = english
    Ref $greeting = hello world
 binding 1
    Ref $language = french
    Ref $greeting = bonjour le monde
```

Part 2: Plans

Now that we have a basic understanding of how beliefs are modelled in GORITE, we will introduce the contextualization of plans, which is at the heart of BDI. In this example, Paul will have a single greetings goal (this time called "greet") modelled as a plan. When a plan is activated, its context() method and precedence() method (if present) are called before plan execution begins (i.e. invocation of its execute() method). In Paul's case, the context method will construct a query that will generate, for a given planet, the languages that he knows about and a translation of "hello world" for each language. In the previous example, where we queried the relation directly, bindings were made accessible through a collection (Vector) of carrier (Ref) objects. In the case of contextualized plans, the bindings are made available as local Ref instances in dynamically generated plan instances – each binding results in a new plan instance. The plan instances are added to the applicable set and the execution ordering can be manipulated through the specification of a precedence value. BDI execution semantics, as explained in Version 3, apply.

The code for the Greet goal is shown below. $ is used to prefix Ref instances; this is just a syntactic convention that we use to distinguish between Ref "variables" and normal Java variables. Also note that precedence could have been modelled as an element of the Greetings relation.

Greet.java:

```java
package chapter2.greetings.v4b;
import com.intendico.gorite.*;
import com.intendico.data.*;

public class Greet extends Plan {

  // name of the goal
  public static final String GREET = "greet";

  Relation can_greet;
  Ref $language = new Ref ( "$language" );
  Ref $greeting = new Ref( "$greeting" );

  public Greet(Relation g) {
    super( GREET );
    can_greet = g;
  }

  public Query context( Data d ) throws Exception {
    String planet = (String) d.getValue( "planet" );
    return can_greet.get(
      new Object [] { planet, $language, $greeting } );
  }

  public int precedence( Data d ) {

    String planet = (String) d.getValue( "planet" );
    if ( planet.equals( "earth" )) {
      if ( "english".equals( $language.get() ) )
        return 7;
      if ( "french".equals( $language.get() ) )
        return 6;
      return 5;
    }
    return 5;
  }

  public Goal.States execute( Data d ) {
    String planet = (String) d.getValue( "planet" );
    System.err.println( "Greet: Landed on "+planet );
    System.err.println( "Greet: "+$greeting.get() );
```

```
    // The spacecraft has landed in Antarctica and the penguins
    // don't understand ...
    System.err.println( "Greet: no response" );
    return Goal.States.FAILED;
  }
}
```

As with the previous example, each `Alien` instance creates and populates a
greetings relation. This is then made available to its `Greet` goal through the goal
constructor. Goal execution is initiated through the `greet()` method which also
constructs the data context for the goal execution and adds to it the planet on
which the spacecraft has landed.

Alien.java:

```
package chapter2.greetings.v4b;
import com.intendico.gorite.*;
import com.intendico.data.*;

public class Alien extends Performer {

  Relation can_greet;

  // allow c[] to be declared as a raw (ie unparameterised) type
  @SuppressWarnings( "rawtypes" )

  // constructor
  Alien( String name ) throws Exception {
    super( name );
    // Beliefs about languages spoken
    // tuples are of the form <planet,language,greeting>
    Class c[] = { String.class, String.class, String.class };
    can_greet = new Relation( "can greet", c );

    // Set greetings
    can_greet.add( "mars", "martian", "VDREW^% ^%FD$^" );
    can_greet.add( "earth", "english", "hello world" );
    can_greet.add( "earth", "french", "bonjour le monde" );

    // Add goals
    addGoal( new Greet(can_greet ) );
  }

  // A method to initiate a greeting goal
  public boolean greet( String planet ) {
    // Greet the people
    Data data = new Data();
    data.setValue( "planet", planet );
    return performGoal( new BDIGoal(Greet.GREET),"Greet",data );
  }
}
```

Finally, the `main()` method for the application creates an `Alien` instance and
invokes its `greet()` method with the planet on which the spacecraft has landed
as its parameter.

Main.java:

```
package chapter2.greetings.v4b;

public class Main {
  public static void main( String [] args ) throws Exception {
    Alien paul = new Alien( "Paul" );
    paul.greet( "earth" );
  }
}
```

The following output is produced when the application is run:

```
Greet: Landed on earth
Greet: hello world
Greet: no response
Greet: Landed on earth
Greet: bonjour le monde
Greet: no response
```

Note that the two plan instances in the applicable set are executed in order of precedence and as the first plan fails, the second plan is then tried (and also fails).

Chapter 3
Process Modelling

In the previous chapter, all the examples involved a performer achieving a single goal that had no substructure and resulted in a simple greeting action being performed. We did allow the performer to choose a goal from multiple goals in the applicable set (version 3) and to dynamically customize a goal through the use of the context() method (version 4, part 2), but in both cases, the goal itself had no sub-structure.

In this regard, the goals of the previous chapter are akin to very simple plans in traditional BDI frameworks – they would map to plans that perform a basic action that either succeeds or fails. Clearly, useful agents need to engage in more complex behaviour than just the performance of independent, basic actions. This complexity has three key aspects:

Procedural complexity. Procedural complexity arises through the procedural composition of basic activities using the typical procedural constructs of sequence, selection, repetition and concurrency. For example, a manufacturing cell performer may have "make a meter box" as a basic action. This basic action can then form the basis for more complex actions such as "make a batch of meter boxes" (repetition) or "make two batches of meter boxes in parallel" (repetition and concurrency). In traditional BDI frameworks, this complexity is realized through constructs provided by the plan language. In some frameworks, the plan language is a superset of the host language so that developers have access to both plan language constructs and host language programming constructs.

Goal complexity. Goal complexity refers to the fact that goal achievement will normally involve the achievement of sub-goals. For example, making a meter box, as we shall see later in this chapter, involves the loading of components into a jig, the transfer of the jig to a joining station where the components are screwed together, the transfer of the joined assembly to an unloading station and then the unloading of the assembly. These actions can be viewed as sub-goals of the "make a meter box" goal. Likewise, "make a meter box" can be viewed as a sub-goal of the "make a batch of meter boxes" goal, which in turn can be viewed as a sub-goal of the "make two batches of meter boxes in parallel" goal. In traditional BDI frameworks, goal complexity is realized through the procedural embedding of sub-goal achievement within plans.

D. Jarvis et al.: Multiagent Systems and Applications, ISRL 46, pp. 31–58.
DOI: 10.1007/978-3-642-33320-0_3   © Springer-Verlag Berlin Heidelberg 2013

Organisational complexity. This arises because complex activity inevitably requires the involvement of multiple agents with specific capabilities. For example, the "make meter box" goal will ultimately involve three separate agents to perform loading and unloading, jig transfer and joining. The behaviours of those agents need to be coordinated at a procedural level and information relating to the overall goal achievement shared and updated. Organisational complexity has been, until the advent of JACK Teams, outside the scope of BDI frameworks. Traditionally, no support has been provided by BDI frameworks to address organizational complexity, with developers having to build and manage the organizational structures required by each application. JACK Teams introduced the notion of a team as being a BDI entity in its own right. The members of a team are BDI entities, so teams can have sub-teams and so on. JACK Teams was formulated as an extension to the traditional BDI model, so consequently it preserves the traditional approaches to both procedural and goal complexity.

While GORITE models team behaviour in a similar way to JACK Teams, it adopts a quite different approach to the issues of procedural and goal complexity, combining both in the form of explicit process models, which are also called goal hierarchies. These models are constructed from objects derived from the base class `Goal` and do not employ a separate plan language. The behavioural aspects of organisational complexity are supported seamlessly in a process model through the use of `TeamGoal` instances. However, the focus of this chapter will be on how process models can be used to manage procedural complexity and goal complexity – organizational complexity is deferred until Chapter 5.

In the remainder of this chapter, we first present an overview of the GORITE modelling classes. We then discuss process modelling in more detail using two case studies – an assembly cell and a sensor network. Both of these systems have an organizational aspect to them; the assembly cell consists of two robots and a rotating table while the sensor network consists of supervisory nodes and sensing nodes. However, as noted above, the organizational aspect of the systems is deferred to Chapter 5. In this chapter, we model the systems as single performers. Also, as we want to focus on process models, we do not develop complete applications as in the previous chapter, but instead focus on the overall process model definition. As a consequence, we do not provide definitions for constants and behaviours for goals are kept as simple as possible. However, complete examples are available from the GORITE web site.

3.1 The Modelling Classes

Process models in GORITE are hierarchical structures composed of goals[1]. These goals fall into two categories – behaviour goals and control overlay goals. The primary focus of behaviour goals is to specify the actual behaviours that will result in the achievement of the process model. In this sense, they form the leaf nodes of the process model and do not contain sub-goals. Control overlay goals, on the

---

[1] Throughout the remainder of this book, we will use the terms *process model, goal, plan* and *goal hierarchy* interchangeably.

other hand, specify the goal decomposition for the process model and the ways in which the sub-goals are to be executed. Although BDI goals (and team goals) don't detail how a behaviour is to be achieved, they do specify the name of the goal that is to be achieved and they appear as leaf nodes in a process model. Consequently, we classify both BDI goals and team goals as behaviour goals and not as control overlay goals.

The basic GORITE goal classes are summarized in the concept map of Figure 3.1:

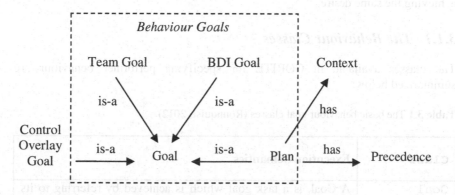

Fig. 3.1 The basic GORITE goal classes[2] (Rönnquist, 2012). The control overlay goals are not elaborated.

In Figure 3.1, the Plan class appears as a behaviour goal. In this capacity, a plan is a goal with context and precedence and performs the same role in a process model as a goal – that is, as a leaf node in a process model and with behaviour specified using an overridden execute() method. The Plan class can also be used as a control overlay goal, in which case no behaviour is specified, but it has sub-goals. When used in this manner, it will normally be as the root node of a process model.

Given the central role that goals play in the modeling of behaviour, it is important to understand what is meant by a goal in GORITE. Conceptually, a goal is the name of a desire, that is an objective or situation that a performer would like to achieve. A goal also provides a definition of how that desire might be achieved. This may involve further definition of sub-goals to elaborate a process for goal achievement. When a performer has committed to the achievement of a particular goal, it becomes an intention. In this book, we generally refer to goals with respect to their execution, so we will often use the term intention instead of goal.

---

[2] Figure 3.1 is not strictly correct in that a goal can have sub-goals. However, this is a design convenience for the control overlay goals of Table 3.3 that extend Goal and can have sub-goals. When an instance of the Goal class is used in a process model, it does not have sub-goals. A plan will often have sub-goals but in that case, the plan is acting as a control overlay goal and not as a behaviour goal.

The above concept of a goal as a named process definition is realised in GORITE through the `Goal` class and its sub-classes. When a definition is executed (i.e. the performer makes an actual attempt to achieve the goal in that way), new objects of type `Goal.Instance` are created to be the actual representation of the intention. Basically, a `Goal` object represents an idea of achieving a goal, whereas a corresponding `Goal.Instance` object represents an actual attempt to achieve that goal. As we have seen with the discussion of BDI goal execution in Chapter 2, goals are characterised by their names. In this regard, the performer interprets goals with the same name as being different ways of achieving the same desire.

3.1.1 The Behaviour Classes

The classes available in GORITE for specifying performer behaviour are summarized below:

Table 3.1 The basic behaviour goal classes (Rönnquist, 2012)

Class	Execution Semantics
Goal	A Goal, is a task goal which is achieved by referring to its `execute()` method. This is Java code, and returns an execution state, which is one of PASSED, FAILED, STOPPED or BLOCKED. Although theoretically a goal is capable of having sub-goals, in practice, it does not.
Plan	A plan is a goal that implements context and precedence. It is achieved in the same way as a goal if its `execute()` method is overridden. Alternatively, a plan can have sub-goals, as explained in Table 3.3, in which any `execute()` method is ignored.
BDIGoal	Generally, a BDI goal is achieved by selecting a goal hierarchy for the nominated goal and executing that hierarchy. If the hierarchy fails, other applicable hierarchies are then considered for execution. The goal succeeds immediately a hierarchy succeeds. The goal fails if all alternatives fail.
AnyGoal	An any goal is achieved like a BDI Goal, but by pursuing all applicable hierarchies in parallel. The goal succeeds/fails immediately a hierarchy succeeds/fails.
AllGoal	An all goal is achieved like a BDI goal, but by pursuing all applicable hierarchies in parallel. The goal succeeds when all hierarchies have succeeded; it fails immediately a hierarchy fails.
TeamGoal	A team goal is a goal directed to a team member; a team member may be an individual agent or a team of agents. The team member is asked to perform the goal as a BDI goal.

As indicated in the previous section, there are two aspects to realizing behaviour in GORITE –

1. the naming of the behaviour to be realised and
2. the detailing of how the named behaviour is to be realised

As we have seen in Chapter 2, in the case of goals and plans these two aspects are not separated – a goal or plan instance has a name and by virtue of an execute() method that can be overridden, a behaviour realization. However, in the case of BDI goals and team goals, the two aspects are decoupled. The name of the behaviour to be achieved is specified by the goal name, but the realization is specified in separate goals/process models that have the same name as the BDI or team goal. This leads to two types of goal execution semantics in GORITE – task semantics and BDI semantics.

With task semantics, which are exhibited by both goals and plans, goal execution is achieved by invoking the goal's execute() method. In contrast, both team goals and BDI goals employ BDI semantics. With BDI semantics, all goal hierarchies that are applicable to the current execution context are considered and incorporated into what is called the applicable set. If the goal hierarchies in the applicable set have been assigned a precedence value, this will be used to rank the hierarchies within the set. Goals within the set are then executed sequentially until a goal succeeds, at which point the BDI/team goal succeeds.

If the applicable set contains multiple plans with the same precedence, plans are attempted in the order in which they appear in the set. If desired, the choice between plans of the same precedence can be made randomly. To achieve this behaviour, a plan choice strategy needs to be specified for the goal in question. This is achieved with the following method:

void Performer.setPlanChoice(String *bname*, Object choice_*strategy*)

bname is the name of the BDI goal and choice_strategy, in this case, is an object of type java.util.Random.

If all members of the applicable set fail, then the BDI/team goal fails. Furthermore, when a goal in the applicable set fails, the set is regenerated, as the conditions for applicability may have changed during the execution of the failed goal.

If finer control over applicable set execution is required, then a goal can be provided to determine which plan in the applicable set is to be attempted next. As with random plan selection, this goal is bound to a BDI goal using the setPlanChoice() method, However, in this case, choice_strategy will reference the name of the goal that is to be used for plan selection.

When a BDI goal is encountered in a process model by the executor, it will determine, via the

Object Performer.getPlanChoice(String *bname*)

method, whether or not one or other of the plan choice strategies described above (random selection or explicit reasoning) applies to bname. If the explicit reasoning strategy is to be employed, the executor will create a PlanChoice Goal instance to manage the selection process. This goal makes available to the user-provided selection goal a new data context that contains the following data elements:

Table 3.2 Data elements available to the user-provided plan that chooses a goal from the applicable set

Name	Description
"options"	The applicable set
"failed"	The set of currently failed plans
"data"	The data context as presented to the BDI goal
"choice"	The chosen plan

The reasoning performed by the user-provided goal is then informed by the first three data elements in Table 3.2. The result of the reasoning process (the selected plan) is stored in the "choice" data element. Examples illustrating the use of both plan choice strategies are available on the GORITE website.

3.1.2 The Control Overlay Classes

The essence of GORITE is that instances of the behaviour classes described in the previous section can be incorporated into process models. This is achieved through the control overlay classes; their purpose is to provide the procedural framework for goal execution, but using goals instead of a separate plan language. The classes are summarized in Table 3.3.

The Plan class is included in both Tables 3.1 and 3.3 because it can be employed both as a basic modelling element (behaviour can be specified by overriding its execute() method) and as a control element (it can have sub-goals, which are executed in sequence as in a sequence goal).

Note that the break effect of an end goal does not escape its enclosing plan/goal. Thus, it has to occur lexically within the sub goal structure of its associated loop goal. This is exactly the same as a "break" statement within a repetition block in Java. In contrast, the control effect of a control goal does escape its enclosing plan/goal and propagates out to the nearest enclosing parallel goal execution. This is notionally the same as an exception in Java, with parallel goals (including team goals) corresponding to enclosing try-catch blocks.

All of the control overlay goals can have sub-goals and can thus act as non-leaf nodes in a process model/goal hierarchy. In addition, all of the goals, with the exception of the parallel goal, can be created without sub-goals and become leaf nodes. In this case the goal's execute() method is overridden to provide the

Table 3.3 The control overlay classes (Rönnquist, 2012)

Class	Execution Semantics
SequenceGoal	A sequence goal is achieved by achieving its sub-goals in sequence. If a sub-goal fails, the goal immediately fails. The goal succeeds if all its sub-goals succeed.
ConditionGoal	A condition goal is achieved by achieving its sub-goals in sequence. The goal succeeds immediately a sub-goal succeeds. It fails if all its sub-goals fail.
FailGoal	A fail goal is achieved by achieving its sub-goals in sequence. It succeeds when a sub-goal fails. The goal fails if all its sub-goals succeed.
LoopGoal	A loop goal is achieved by achieving its sub-goals as a sequence goal until one of the sub-goal hierarchies achieves an end goal. A loop goal fails immediately a sub-goal fails, and succeeds when an inner end goal succeeds.
EndGoal	If all the sub-goals of an end goal succeed when attempted in sequence, then the end goal succeeds and breaks its enclosing loop goal. If a sub-goal fails, the end goal succeeds, but without breaking the loop. An end goal never fails.
ParallelGoal	A parallel goal is achieved by achieving its sub-goals in parallel. The goal succeeds when all branches have succeeded. It fails when a sub-goal fails. Further, if a sub-goal fails, all other sub-goals are cancelled.
RepeatGoal	A repeat goal replicates its sub-goals n times, where n is the number of elements in the goal's control element. This element is a multi-valued data element in the data context that is made available to the repeat goal at construction time. Each replica gets a separate data context focusing on one of the values of the control element. The replicas are then processed as parallel branches. A repeat goal succeeds when all branches have succeeded. If a branch fails, then the goal fails and all other branches are cancelled.
ControlGoal	A control goal has control semantics similar to an end goal, but for parallel goal executions (i.e. a ParallelGoal or a RepeatGoal). Its sub-goals are achieved in sequence, and if all succeed, then the enclosing parallel execution succeeds and all branches are cancelled.
Plan	Plan extends SequenceGoal, so if sub-goals are present in a plan, the plan is achieved in the same way as a sequence goal and any execute() method is ignored.

behaviour that would otherwise have been provided as explicit sub-goals. Note
that in both cases, the execution semantics detailed in Table 3.3 are provided by
the GORITE infrastructure independently of a particular goal's execute()
method.

3.2 The Meter Box Cell

For our first example in this chapter, we will consider the manufacturing control
application described in (Jarvis et al, 2006). It involved the assembly of meter
boxes which consist of either two components, A and B, or three components, A,
B and C. In this case, we will only be concerned with the assembly of boxes with
two components, which we will refer to as ABs. The assembly cell, when
configured for the assembly of ABs, consisted of 3 machines – a pick and place
robot (referred to as Fanuc), a screwdriver robot (referred to as Hirata) and a
rotating table. The rotating table had two jigs located diametrically opposite each
other, so that if desired, two boxes could be assembled concurrently. The Hirata
and Fanuc were positioned at opposite sides of the table. The Fanuc was
responsible for the loading of A's and B's into the jigs and the unloading of
assembled AB's. The Hirata was responsible for the joining of As to Bs. A
schematic layout of the cell is shown below.

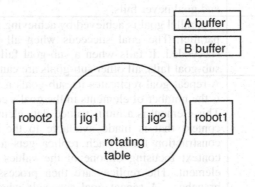

Fig. 3.2 The Meter Box Assembly Cell (Jarvis et al., 2006) configured for the assembly of
ABs

A picture of the actual cell (now decommissioned) can be seen in Figure 3.3.

An Omron PLC (programmable logic controller) provided basic control
functionality (control of table motion, control of jig opening and closing,
triggering of Fanuc and Hirata actions). An execution system employing this
functionality was developed using JACK Teams (AOS Group, 2012).

In this chapter, we will replicate the key behaviour of that system using
GORITE. However, we will make one key simplifying assumption – that the
assembly behaviour is achieved by a single performer and not a team of
performers. That way, we will not be distracted by organizational issues – these

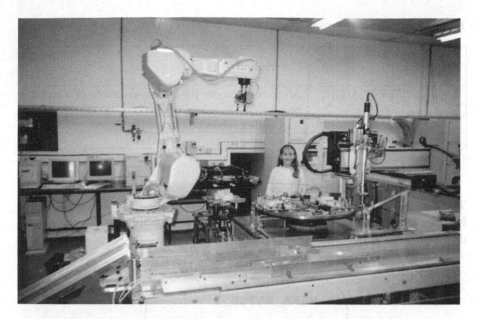

Fig. 3.3 The Meter Box Assembly Cell. The Fanuc robot and the Flipper unit are to the right of Mairi Jarvis. The rotating table is in front of her and the Hirata robot is to her left. The Flipper unit was not used in the making of ABs.

will be addressed in Chapter 5. Also, we are interested only in simulated operation, not the actual physical operation of the cell as in (Jarvis et al., 2006). As a consequence, operations are modelled as print statements and time delays.

In this chapter, we will develop a process model that defines how the meter box cell can make a batch of ABs utilising both jigs. The process model will be developed incrementally as three distinct versions:

> Version 1:   Making a single meter box
> Version 2:   Making a batch of meter boxes in a single jig
> Version 3:   Making a batch of meter boxes using two jigs

The reason for having three versions is primarily pedagogical as it enables the key control goals for sequence, repetition and concurrency to be introduced in a progressive manner. However, it also demonstrates that phased implementation is just as applicable to multi-agent systems development as it is to conventional software development.

3.2.1 Version 1: Making a Single Meter Box

A meter box consists of 2 components – A and B. To make a meter box, a jig needs to be assigned to the task, the box assembled and then the jig released:

Table 3.4 The high level steps required to make a meter box

Step	Goal	Description
1	RESERVE_JIG	Allocate an empty jig for the assembly of a meter box
2	MAKE_METERBOX	Assemble a meter box in the jig
3	RELEASE_JIG	Release the jig for further work

These steps represent the top level decomposition of a MAKE_BATCH goal. To distinguish between versions, we will append the version number to MAKE_BATCH. Thus, diagrammatically, we have the following top level goal decomposition[3] for our first version, MAKE_BATCH_1:

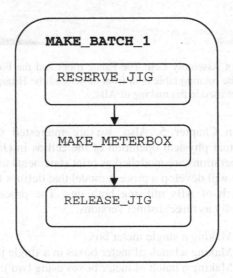

Fig. 3.4 The first level decomposition of the MAKE_BATCH_1 goal

MAKE_METERBOX is broken down into the following tasks, which are executed in sequence:

---

[3] The representation used in Figure 3.4 was chosen to emphasise the goal decomposition. UML Activity Diagrams (Fowler, 2003) or BPMN (OMG, 2011) could have been subverted for our purposes, but both approaches are based on flowchart notations which do not capture decomposition well. Also, we wanted to emphasise that all nodes in Figure 3.4 are goals.

Table 3.5 The high level steps required to assemble a meter box in a jig

Step	Goal	Description
1	MOVE_TO_LOADER	Move the allocated jig to the loading station
2	LOAD_A	Load an A into the jig
3	LOAD_B	Load a B into the jig
4	MOVE_TO_JOINER	Move the jig containing the components to the joining station
5	JOIN_AB	Join the two components
6	MOVE_TO_LOADER	Move the jig to the loading/unloading station
7	UNLOAD_AB	Unload the assembled meter box from the jig

Diagramatically, the goal decomposition is as shown below:

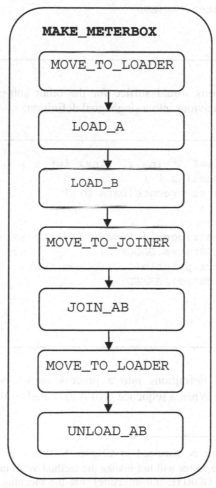

Fig. 3.5 Decomposition of the MAKE_METERBOX goal

We begin our discussion with the MAKE_METERBOX goal. In terms of a process model, the individual sub-goals can be modelled (for now) as `Goal` instances as in the previous chapter. For example, LOAD_A could be modelled as follows[4]:

```
new Goal( LOAD_A ) {
  public States execute(Data d) {
    // TimeTrigger is defined in com.intendico.gorite.addon
    // isPending() sets an alarm for 5 seconds (5000 msec) in
    // the future. Until it goes off, BLOCKED is returned.
    if ( TimeTrigger.isPending( d, "deadline", 5000 ) )
      return States.BLOCKED;
    System.err.println("A loaded in jig");
    return States.PASSED;
  }
}
```

Similar definitions would suffice for the other sub-goals, so we choose to encapsulate the behaviour into a single goal definition:

```
Goal step( final String s, final int t ) {
  return new Goal( s ) {
    public States execute(Data d) {
      // perform the step. s and t are declared final, and
      // are therefore in scope.
      if ( TimeTrigger.isPending( d, "deadline", t*1000 ) )
        return States.BLOCKED;
      System.err.println("Performed "+s);
      return States.PASSED;
    }
  };
}
```

To turn the goal definitions into a process model, we employ a GORITE `SequenceGoal`. When a sequence goal is executed by the performer's executor,

---

[4] Blocking behaviour is discussed in Chapter 5. When an execute() method returns BLOCKED, the executor will not invoke the method again until it has been notified (in this case, by the GORITE infrastructure) that the blocking condition may no longer apply.

each of its sub-goals are executed in sequence. If all the sub-goals succeed, then the sequence goal succeeds. If a sub-goal fails, then the sequence goal immediately fails.

A sequence goal has the following constructor:

```
SequenceGoal( String name, Goal[] list )
```

where name is the name of the sequence goal instance and list is an array of sub-goal instances. The elements of list can be of any type that sub-classes Goal, including SequenceGoal – note that both MAKE_BATCH_1 and MAKE_METERBOX will be modelled as sequence goals and that MAKE_METERBOX will be a sub-goal of MAKE_BATCH_1.

As we have seen in Chapter 2, a goal class can be defined as a normal Java class or it can be defined at the point of use via an anonymous class. With the definition of the step() method, we have introduced a third option, whereby a method returns an instance of the required class. Within the step() method, the class was defined inline. In this book, we will use this third approach to model re-usable goal structures. Thus the MAKE_METERBOX goal is returned by the makeMeterbox() method:

```
Goal makeMeterbox() {
  return new SequenceGoal( MAKE_METERBOX, new Goal[] {
    step( MOVE_TO_LOADER, 1 ),
    step( LOAD_A, 2 ),
    step( LOAD_B, 2 ),
    step( MOVE_TO_JOINER, 1 ),
    step( JOIN_AB, 2 ),
    step( MOVE_TO_LOADER, 1 ),
    step( UNLOAD_AB, 1 ),
  });
}
```

The top level goal, MAKE_BATCH_1, is then defined in a similar fashion to MAKE_METERBOX, with makeBatch1() returning an instance of a MAKE_BATCH_1 goal:

```
Goal makeBatch1() {
 return new SequenceGoal( MAKE_BATCH_1, new Goal[] {
   step( RESERVE_JIG, 0 ),
   makeMeterbox(),
   step( RELEASE_JIG, 0 ),
  });
}
```

The MAKE_BATCH_1 goal can then be added to the performer's default capability by including the following statement in its constructor:

```
   addGoal( makeBatch1() );
```

3.2.2 Version 2: Making a Batch of Meter Boxes in a Single jig

Having defined a process model for making a single meter box, we will now reuse that model to develop a process model to make a batch of meter boxes in a single jig. The overall structure is shown in Figure 3.6 below.

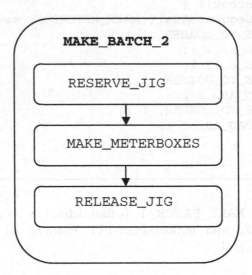

Fig. 3.6 Decomposition of the MAKE_BATCH_2 goal

In Figure 3.6, we have introduced two new goals – MAKE_BATCH_2 and MAKE_METERBOXES. The intent is that MAKE_METERBOXES will make multiple meter boxes, so it will have MAKE_METERBOX as a sub-goal. MAKE_METERBOXES will be modelled as a loop goal that has MAKE_METERBOX as its body and TRACK_ORDER as its end goal; TRACK_ORDER will succeed when the required number of meter boxes have been made. A suitable process model expression could look like the following:

```
new LoopGoal( MAKE_METERBOXES, new Goal [] {
  makeMeterbox(),
  // trackOrder() returns an EndGoal
  trackOrder(),
})
```

This expression will be returned by a makeMeterboxes() method. We assume that the number of boxes to be assembled is provided as an entry in the data context for the goal execution, so trackOrder() is defined as follows:

```
EndGoal trackOrder() {
  return new EndGoal( TRACK_ORDER ){
    public Goal.States execute( Data d ) {
      // An order object specifies the number of boxes required
      //(required) and tracks of the number of boxes that have
      // been assembled (count).
      Order o = (Order) d.getValue( ORDER );
      o.count++;
      System.err.println( "Made "+o.count+" meter boxes " );
      // has the goal been achieved?
      if ( o.count == o.required ) {
        System.err.println( "Order filled" );
        return Goal.States.PASSED;
      }
      return Goal.States.FAILED;
    }
  };
}
```

The top level goal, MAKE_BATCH_2, is then defined in a similar fashion to the MAKE_BATCH_1 goal and is returned by the makeBatch2() method:

```
Goal makeBatch2() {
  return new SequenceGoal( MAKE_BATCH_2, new Goal[] {
    step( RESERVE_JIG, 0 ),
    makeMeterboxes(),
    step(RELEASE_JIG, 0 ),
  });
}
```

3.2.3 Version 3: Making a Batch of Meter Boxes Using Two jigs

With our current process model, only a single jig is used even if both jigs are available. To remedy this situation, we will allow an order to be filled using both jigs concurrently. There is now a synchronization problem that needs to be addressed, in that the table must not move while either of the two stations are busy. We will ignore the problem for now, but will address it in Chapter 5, when we model the meter box cell as a team.

From a modelling perspective, GORITE distinguishes between two different ways in which concurrent intentions arise:

1. a single intention is split into multiple branches that are then executed in parallel.
2. separate intentions are executed in parallel.

These are realized using the RepeatGoal and ParallelGoal respectively. Parallel goals will be discussed in the sensor network example that follows – for now, we will focus on repeat goals and illustrate their use by splitting the make meterboxes intention into two branches, one for each jig.

With a repeat goal, one element of the data context is specified to be the *control element* for the goal. As we shall see later, this binding takes place at goal construction. The number of values in the control element determines the number of concurrent branches that will be created for the repeat goal and each branch is bound to a particular value of the data element. In our example, the multi-valued data element JIG will be the control element. Prior to goal execution, JIG will be assigned the two values JIG1 and JIG2 as follows:

```
d.setValue( JIG, JIG1 );
d.setValue( JIG, JIG2 );
```

so two branches will be created. Within a branch, a reference to JIG will result in the appropriate value being accessed. For example, the statement

```
// d is the data context
String j = (String) d.getValue( JIG );
```

will return JIG1 from branch 1 and JIG2 from branch 2.

In this particular case, JIG is the only element in the data context that will need to assume different values for different branches. If this was not the case, then one could either

1. make the control element values branch configuration objects, where a branch configuration object is a container for all branch-varying data or
2. make the control element values the branch indices (0 .. n). These indices are then used to access branch-varying data modelled as multiple multi-valued data items.

In the second case, branch data is accessed using DataElement.get():

```
// BRANCH is the control element, d is the data context
// get the branch index.
int b = ((Integer) d.getValue(BRANCH)).intValue();
// get the jig name for this branch.
// DataContext.find(JIG) returns the DataElement instance named
// JIG.
// DataElement.get(b) returns the bth value of the data element.
String j = (String) d.find(JIG).get(b);
```

BRANCH will have previously been assigned two values as follows:

```
d.setValue( BRANCH, 0 );
d.setValue( BRANCH, 1 );
```

While a branch can access a multi-valued data element directly, the intended practice is that if a branch needs to manage its own value of a multi-valued data item, the branch "shadows" the item. For example, our assembly application will need to record the total number of boxes that have been made, but it may make sense for each branch to also keep track of the number that they have individually produced. This situation could be modelled by having an element called TOTAL that is then shadowed in each branch. A branch will then update its local shadowed value as well as the global value. On completion of the repeat goal, the shadowed values can then become additional values of the global element. Note that data element access is thread safe, so multiple branches can safely access shared elements without the need for explicit synchronization.

A repeat goal is specified as a list of sub-goals using the following constructor:

```
RepeatGoal(String cname, String gname, Goal[] list )
```

cname is the name of the control element and gname is the name of the repeat goal. The list of sub-goals constitute the branches for the repeat goal and are progressed by the executor concurrently. Normal BDI execution semantics apply – that is if a branch fails, then the remaining branches are cancelled and the repeat goal fails and if all branches succeed, the repeat goal succeeds. If finer control over branch execution is required, then a ControlGoal can be employed. The sub-goals of a control goal are achieved in sequence, and if all succeed, then the enclosing parallel execution succeeds immediately and all remaining branches are cancelled.

The overall structure for the making of meter boxes in two jigs is shown in Figure 3.7:

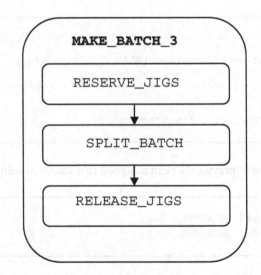

Fig. 3.7 Decomposition of the MAKE_BATCH_3 goal

As before, our interest is not in jig management, so RESERVE_JIGS and RELEASE_JIGS are ignored – they will be covered in Chapter 5. Note that as SPLIT_BATCH is a repeat goal, it will use two jigs if two jigs are available. However, if only one jig is available, we will have a repeat goal with a single branch and only one jig will be used.

As our first attempt, we define a `splitBatch()` method as follows:

```
Goal splitBatch() {
  // The strategy that we are employing is to use all available
  // jigs (2) for assembly. Each jig will have its own branch in
  // a RepeatGoal. JIG is the control element.
  return new RepeatGoal( JIG, SPLIT_BATCH, new Goal [] {
    makeMeterboxes()
  });
}
```

The `MAKE_METERBOXES` goal returned by `makeMeterBoxes()` was defined in Version 2 as

```
new LoopGoal( MAKE_METERBOXES, new Goal [] {
  makeMeterbox(),
  // trackOrder() returns an EndGoal
  trackOrder(),
})
```

The `splitBatch()` method can then be embedded in a method that returns an instance of a `MAKE_BATCH_3` goal:

```
Goal makeBatch3() {
  return new SequenceGoal( MAKE_BATCH_3, new Goal[] {
    step( RESERVE_JIGS, 0 ),
    splitBatch(),
    step( RELEASE_JIGS, 0 ),
  });
}
```

When `SPLIT_BATCH` is executed, a branch of the repeat goal will succeed when `TRACK_ORDER` succeeds. However, the repeat goal will succeed only when **both** branches have succeeded and an extra box may be made. To force the repeat goal to succeed immediately on the completion of the first branch, we could wrap the branches in a control goal as follows:

```
Goal splitBatch() {
  // The strategy that we are employing is to use all available
  // jigs (2) for assembly. Each jig will have its own branch in
  // a RepeatGoal. JIG is the control element.

  return new RepeatGoal( JIG, SPLIT_BATCH, new Goal [] {
    new ControlGoal("stop on first branch completion",
      new Goal [] {
          makeMeterboxes(),
    })
  });
}
```

However, what will now happen is that when the first branch completes, the
second branch will be cancelled, even though it may be in the middle of making a
meter box. So we can either revert to our original approach and perhaps make an
extra box or not allow a new meter box to be made if the final meter box is being
made. The latter option can be realized by changing the structure of the MAKE_
METERBOXES goal so that testing for order fulfillment is done before a meter box is
made. To achieve this, we replace the functionality of TRACK_ORDER with two
separate goals – CHECK_ORDER and UPDATE_ORDER that are returned by
checkOrder() and updateOrder() respectively:

```
Goal makeMeterboxes() {
  return new LoopGoal( MAKE_METERBOXES, new Goal [] {
    checkOrder(),
    makeMeterbox(),
    updateOrder(),
  });
}
```

CHECK_ORDER is an EndGoal, so when it succeeds, MAKE_METERBOXES will
immediately succeed. What we now need to do is to stop a branch if either the
requisite number of boxes have been made or if the last box is being made by the
other branch. We do this by introducing an additional field, wip, to the Order
class to keep track of the boxes that are currently being made. checkOrder()
then becomes

```
EndGoal checkOrder() {
  return new EndGoal( CHECK_ORDER ){
    public Goal.States execute( Data d ) {
      Order o = (Order) d.getValue( ORDER );
      // has the goal been achieved?
```

```
    if ( o.count >= o.required ) {
        System.err.println( "Order filled" );
        return Goal.States.PASSED;
    }
    if ( o.count + o.wip >= o.required ) {
        System.err.println( "Order about to be filled" );
        return Goal.States.PASSED;
    }
    // make another box
    o.wip++;
    return Goal.States.FAILED;
    }
  };
}
```

updateOrder() decrements wip and increments count:

```
Goal updateOrder() {
  return new Goal( UPDATE_ORDER ){
    public Goal.States execute( Data d ) {
      Order o = (Order) d.getValue( ORDER );
      o.wip--;
      o.count++;
      System.err.println( "Made "+o.count+" meter boxes" );
      return Goal.States.PASSED;
    }
  };
}
```

3.3 Sensor Networks

In this example, as with the meter box cell, we focus on process model construction without being sidetracked by organisational issues. Therefore, we choose to model our sensor network as a single performer and not a team of performers – it will be modelled as a team of performers in Chapter 5. Our sensing agent can sense temperature and humidity and so has goals to realise this functionality – READ_TEMPERATURE and READ_HUMIDITY. However, the humidity sensor is unreliable and can fail if the power falls below a certain level. The sensing agent thus needs to include a capability for dealing with this.

We will develop three related models:

1. An initial model in which neither sensors nor agents fail
2. A model that detects sensor failure
3. A model that monitors power level

In models 2 and 3, detection of a failure will result in the top-level goal failing. In Chapter 5, fault handling will be added – this will involve the dynamic modification of the sensing team structure.

3.3.1 Version 1: Fault-Free Reading

We begin with a baseline example – no failures occur in either the sensors or the agent. The agent reads the sensors every 5 seconds and this continues forever. Reading is modelled as a loop goal without an associated end goal.

The FAULT_FREE_READING goal is initially defined in the following expression:

```
new LoopGoal( FAULT_FREE_READING, new Goal [] {
  new Goal( READ_TEMPERATURE ) {
    public Goal.States execute( Data d ) {
      // Temperature is constant
      System.err.println( "T = 25" );
      return Goal.States.PASSED;
    }
  },
  new Goal( READ_HUMIDITY ) {
    public Goal.States execute( Data d ) {
      // Humidity is constant
      System.err.println( "H = 50" );
      return Goal.States.PASSED;
    }
  },
  new Goal( PAUSE ) {
    public Goal.States execute(Data d) {
      // delay interval is in seconds - convert it to msecs.
      Integer t = (Integer) d.getValue( INTERVAL );
      if ( TimeTrigger.isPending( d,"alarm",t.intValue()*1000 ) )
        return Goal.States.BLOCKED;
      return Goal.States.PASSED;
    }
  },
  // Read for ever. If required, an end goal would go here.
}
```

Pausing is modelled as a delay using the TimeTrigger class, as in the meter box examples of the previous section. The delay interval is extracted from the data context for the goal execution – the setting of its value is not shown. In the meter box examples, process models were constructed through the use of goal construction methods, as in Version 3:

```
Goal makeMeterboxes() {
  return new LoopGoal( MAKE_METERBOXES, new Goal [] {
    checkOrder(),
    makeMeterbox(),
    updateOrder(),
  });
}
```

However, in the current example, we have chosen to define the sub-goals explicitly within the surrounding control overlay goal, FAULT_FREE_READING. Which approach is used is a question of personal preference. In this case, we would argue that because of the simplicity of the sub-goal behaviours, using goal construction methods would not improve on readability. Also, as the purpose of this section is to explore how failure impacts on process model structure, the reading goals will not be reused.

This example captures what Coplien and Bjørnvig (2010) refer to as the "sunny day" scenario in their discussion of use cases – that is, the scenario where everything goes according to plan. Scenario branching for this application and its impact on process model design is introduced in subsequent versions. One could argue that the artefacts of process model design (process models) are conceptually the same as those of conventional requirements analysis (use cases). From this perspective, process models are executable use cases and the subsequent versions in this section are elaborations on this initial sunny day scenario. This represents a different development stance to that taken for the meter box versions of the previous section, which were posed in terms of phased implementation.

3.3.2 Version 2: Sensor Failure

In this example, the humidity sensor will fail after two readings. The sensing agent, referred to as the supervisor, will initiate a software reset of the sensor. If the sensor fails again, then the top-level sensing goal will fail. In Chapter 5, the second sensor failure will be trapped and the failed sensor replaced by a suitably configured sensor, if one is available.

In order to perform the reset, we need to monitor and control sensor state, so we introduce a Controller class. This class will form the basis for the sensor team members in Chapter 5. For now, the supervisor will have two instance variables, hc and tc that are controllers for humidity and temperature respectively. The Controller class is outlined below (refer to the GORITE website for the full implementation):

```
class Controller {

  // Variable definitions / declarations go here
  // Constructor goes here

  public boolean reset(){
    if (resets > 0)
        return false;
    count = 0;
    resets++;
    status = OPERATIONAL;
    return true;
  }

  public boolean read(){
    if ( mode == UNRELIABLE && count == 2 ) {
        status = FAILED;
        reading = 0;
        return false;
    }
    if ( source.equals( TEMPERATURE ))
        reading = 25;
    if ( source.equals( HUMIDITY ))
        reading = 50;
    count++;
    return true;
  }
}
```

Rather than defining separate READ_HUMIDITY and READ_TEMPERATURE
goals as we did in Version 1, we create the goals using a `read()` method:

```
Goal read(String name, final Controller c) {
  return  new Goal( name ){
    public Goal.States execute( Data d ) {
      if ( !c.read() ) {
          System.err.println( c.name+": failed" );
          return Goal.States.FAILED;
      }
      System.err.println(
        c.name+": < "+c.type+","+c.reading+","+c.count+">" );
      return Goal.States.PASSED;
    }
  };
}
```

Sensing is now achieved using the BASIC_READING goal, which is defined as
follows:

```
new LoopGoal(BASIC_READING, new Goal [] {
  read( TEMPERATURE, tc ),
  read( HUMIDITY, hc ),
  new Goal( PAUSE ) {
    public Goal.States execute( Data d ) {
      // delay interval is in seconds - convert it to msecs.
      Integer t = (Integer) d.getValue( INTERVAL );
      if ( TimeTrigger.isPending( d,"alarm",t.intValue()*1000))
        return Goal.States.BLOCKED;
      return Goal.States.PASSED;
    }
  },
  new EndGoal( READING_COMPLETED ) {
    public Goal.States execute( Data d ) {
      // read for ever
      return Goal.States.FAILED;
    }
  }
}
```

If we were to now run this model with the humidity controller configured to operate in unreliable mode, output similar to the following would be produced:

```
t1: < temperature,25,1>
c1: < humidity,50,1>
t1: < temperature,25,2>
c1: < humidity,50,2>
t1: < temperature,25,3>
c1: failed
```

The read humidity goal fails on its third invocation. As this goal is a sub-goal of the BASIC_READING loop goal, BASIC_READING fails and execution stops. If we now decide to handle the failure by initiating a reset and continuing with execution, we can do this through the introduction of a new goal, ROBUST_ READING:

```
new LoopGoal( ROBUST_READING, new Goal [] {
  new EndGoal( MONITOR, new Goal[] {
    basicReading()
  }),
  new Goal( HANDLE_FAULT ) {
    public Goal.States execute( Data d ) {
      return Supervisor.this.handleFault(d);
    }
  },
}
```

What we have done in the ROBUST_READING goal is to wrap the BASIC_
READING goal (returned by basicReading()) in an end goal. If a sensor
doesn't fail, reading continues as expected. However, if the reading goal fails then
the end goal succeeds, the loop is not broken and execution progresses to the fault
handling goal.

The fault handling goal calls handleFault(), which is defined as follows:

```java
public Goal.States handleFault( Data d ) {
   System.err.println(
    "supervisor: handling fault by resetting sensor");
   // assume a single point of failure
   Controller c = ( hc.status == Controller.FAILED ) ? hc : tc;
   if ( c.reset() )
     return Goal.States.PASSED;
   System.err.println(
     "supervisor: resetting of "+c.name+" unsuccessful");
   return Goal.States.FAILED;
}
```

When the sensor is reset successfully, the HANDLE_FAULT goal succeeds and
the ROBUST_READING intention continues. However, when the sensor fails for
a second time, the reset will be unsuccessful and the fault handling goal will fail.
This will then cause the ROBUST_READING goal to fail. The following output
will be produced:

```
t1: < temperature,25,1>
c1: < humidity,50,1>
t1: < temperature,25,2>
c1: < humidity,50,2>
t1: < temperature,25,3>
c1: failed
supervisor: handling fault by resetting sensor
t1: < temperature,25,4>
c1: < humidity,50,1>
t1: < temperature,25,5>
c1: < humidity,50,2>
t1: < temperature,25,6>
c1: failed
supervisor: handling fault by resetting sensor
supervisor: resetting of c1 unsuccessful
```

3.3.3 Version 3: Supervisor Failure

In this example, the supervisor will monitor its power level while it is taking
readings. When the power drops below a particular level, the combined reading
and monitoring behaviour will fail. We choose to model this scenario using a
parallel goal with two branches, one for reading and one for monitoring power:

```
new ParallelGoal( POWER_AWARE_READING, new Goal [] {
  robustReading(),
  new Goal( "monitor power level" ) {
   public Goal.States execute( Data d ) {
     // power is an instance variable of an enclosing class and
     // is therefore in scope
     int t = ((Integer) d.getValue( THRESHOLD )).intValue();
     if ( power > t )
       return Goal.States.STOPPED;
     System.err.println(
        "supervisor: reading stopped - power below threshold);
     return Goal.States.FAILED;
   }
 }
})
```

robustReading() will return an instance of the ROBUST_READING goal defined previously. As with a repeat goal, the semantics of parallel goal execution are that the goal passes if all branches pass and the goal fails if any branch fails. In this instance this is the desired behaviour, so no control goals are required.

Belief modelling is simplistic, with power being modelled as an instance variable rather than as a relation. In the latter case, GORITE infrastructure in the form of the Reflector class could be used to block execution of the power monitoring goal until the power dropped below the threshold value. This refinement will be added in our discussion of beliefs in Chapter 6.

If one would want to trap supervisor failure, then the POWER_AWARE_READING goal could be wrapped in a fail goal as follows:

```
Goal sensorReading() {
   // Plan could be replaced with SequenceGoal as we are not
   // utilising precedence or context
   return new Plan ( SENSOR_READING, new Goal[] {
     new FailGoal ( "trap power failure" , new Goal [] {
        powerAwareReading(),
     }),
     new Goal( "handle power failure" ) {
        public Goal.States execute( Data d ) {
           // fault handling code goes here
           System.err.println( "supervisor: "+
              " notifying base station of loss of power" );
           return Goal.States.PASSED;
        }
     }
   } );
}
```

In this case, failure is most likely unrecoverable, so all that the fault handling code would do is to notify the base station of the situation and return[5].

---

[5] In Chapter 5, supervisor failure will result in the sensors of the failed supervisor being redistributed to other supervisors, as the sensor network will be modelled as a team structure. In this example, the network is modelled as a single agent.

Chapter 4
Situated Action

Situated agents interact with their environment. This environment may be virtual as in the war gaming examples of Chapter 1 or it may be physical as in DaimlerChrysler's P2000+ system (Schild and Bussmann, 2007). In either case, the agent must be able to both perceive and act on its environment. The BDI model provides no direct support for such interaction models, focusing instead on the intermediary process of deliberation, that is the determination of what course of action (intention) should be pursued. This process involves the use of models of the environment constructed by the agent. Whether such modelling is necessary to achieve intelligent behaviour was first questioned by Brooks (Brooks, 1999) and the use of the actual environment as the agent's "model" underpins his subsumption architecture. At the other extreme, we have cognitive reasoning models like OODA (Coram, 2002). These models provide a framework for structuring actions and their interactions according to accepted cognitive processes – for example, observation, orientation, decision and action in the case of OODA, which is summarized in the diagram below.

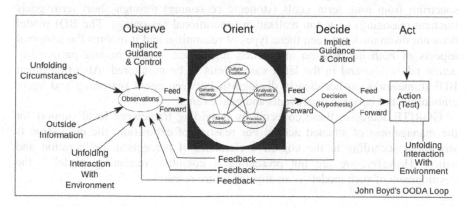

Fig. 4.1 The OODA Loop (Moran, 2008)

While the BDI model, with its focus on the use of internal models (i.e. beliefs) to guide deliberations on intention, provides a counterpoint to behaviour-based approaches such as the subsumption architecture, its relationship to cognitive reasoning models is not so clear. BDI frameworks have been used to implement cognitive reasoning models for agents situated in both virtual and physical

D. Jarvis et al.: Multiagent Systems and Applications, ISRL 46, pp. 59–76.
DOI: 10.1007/978-3-642-33320-0_4   © Springer-Verlag Berlin Heidelberg 2013

environments, as in, for example, (Lui et al., 2002) and (Karim and Heinze, 2005). Also, the cognitive modeling framework, CoJACK (AOS Group, 2012) was implemented using a BDI framework, namely JACK. Cognitive reasoning models provide for elaboration of the deliberation aspect of the BDI model. However such reasoning is explicitly grounded in perception of the environment and action on the environment – activities that are outside the scope of the BDI model. Furthermore, as evidenced by the OODA model, interaction between the deliberation (orient and decide) and observe and act modules is complex and non-sequential.

Whether a cognitive reasoning model such as OODA provides a more appropriate starting point for the development of situated intelligent agent applications than the BDI model is an open question. However, what is clear is that the BDI model, unlike cognitive reasoning models and behaviour-based models, does not have a well-developed concept of interaction with the environment within which an agent is situated. Providing support for such a model so that operational reasoning is integrated with tactical and strategic reasoning is difficult with the traditional BDI execution model, given its underlying representational framework. Of particular importance in this regard is that operational reasoning is firmly grounded in reasoning about both current and future intentions. In responding to a new situation, an agent needs to choose between alternative responses. While this is supported by traditional BDI frameworks, the agent also needs to consider its current intentions and the impact that any new course of action will have on those intentions. In the absence of an explicit and manipulable representation of intention, such reasoning is difficult to support in a generic manner.

Ultimately, reasoning for a situated, intelligent agent needs to span the spectrum from long term goals (strategic reasoning) through short term goals (tactical reasoning) to action realization (operational reasoning). The BDI model does not distinguish between these types of reasoning, and it ignores the temporal aspects of both deliberation and action realization – thought and particularly action takes time and in the latter case, needs to be monitored. Also traditional BDI frameworks do not clearly distinguish between agent reasoning and agent embodiment or between data events and goal events.

GORITE addresses these concerns by providing framework level support for the management of situated action. For reasons of exposition, the discussion is structured according to the cognitive elements of perception, deliberation and action. However, we are not presenting a cognitive reasoning model – the development of such models is an area for future research.

4.1 Deliberation

In traditional BDI frameworks, execution is agent focused – an individual agent determines what plan to perform in order to achieve its current goal. This determination is conducted in the absence of any explicit representation of currently active or future intentions. In GORITE, additional modelling support is provided to enable an agent to

1. better participate in the achievement of larger system-level goals and
2. reason about the progression of multiple concurrent goals

Involvement in system-level goals is supported by GORITE through the use of process models (Chapter 3) and teams (Chapter 5). Support for the concurrent pursuit of multiple goals is provided through the concept of a ToDo Group.

Each performer can maintain a ToDo Group, which is a list of the intentions that it is currently pursuing or has decided to pursue in the near future. Only one intention in a ToDo group is progressed during a time slice[1] – that is the intention that is at the top of the list. However, the ordering of intentions in a ToDo group can be changed through the use of meta-goals. A ToDo group can be bound to a meta-goal using the

```
Performer.addTodoGroup(  String  todo_group,  String
meta_goal )
```

method. If a bound meta-goal exists, the executor will invoke its execute() method at the beginning of every time slice for the ToDo group. The meta-goal is able to access the ToDo group through the data context and re-order its elements if appropriate.

When manipulating a ToDo group, note that ToDo group elements are of type Goal.Instance and not of type Goal. What we haven't told you is that goals are not directly executed in GORITE but rather, that the executor creates an instance object for each goal. Both the instance object and its associated goal object participate in goal execution, but execution is driven by the instance object. Also note that the state of a ToDo group element has two components:

1. the last value returned from an invocation of the execute() method for the goal. This is accessible via the instance object's state member. This is of type Goal.States and will have a value of either STOPPED, BLOCKED or CANCELLED[2].
2. an indication as to whether or not the intention wants to progress. This is accessible via the instance object's data context by invoking the Data.isRunning() method. It returns true or false – true means that the intention wants to progress and false means that it does not.

---

[1] In GORITE, an executor object executes a process model on behalf of the performers that have been assigned to achieve the goals that constitute the process model. The executor traverses the model, instantiates sub-goals and invokes their execute() methods. In GORITE, executors run in separate threads but within each thread, all goals need to be progressed fairly. This is achieved using a concept of focus – each sub-goal is allocated a maximum number of execution cycles. While a sub-goal has focus, the executor will query the sub-goal with respect to the status of its execution. If it has completed, or it is blocked, focus will be shifted to the next active sub-goal.

[2] If PASSED or FAILED had been returned, the goal would no longer be in the ToDo group as would have completed its execution.

Furthermore, note that when an element is added to a ToDo group, `state` is set
to `STOPPED` and the isRunning flag is set to `true`.

As an example of ToDo group reasoning, consider the meta-goal below:

```
new Plan( "percepts meta goal" ) {
  public States execute(Data d) {
    // access the ToDo group
    Performer.TodoGroup todo =
      (Performer.TodoGroup) d.getValue( "todogroup" );

    // access the elements of the ToDo group, which are instances
    // of type Goal.Instance
    System.err.println( "reviewing " + todo.name );
    int found = -1;
    for (int i = 0; i < todo.stack.size(); i++ ) {
      Goal.Instance instance = (Goal.Instance) todo.stack.get(i);
      String g = instance.getGoal().name;
      States s = instance.state;

      // thread name is a concatenation of the head for the goal
      // execution and the goal name
      boolean running =
        instance.data.isRunning( instance.thread_name );
      System.err.println("[" + i + "] Goal " + g + " " + s
        + ( running ? " running" : " monitored" ) );
      if ( "abort".equals( g ) )
        found = i;
      else if ( found == -1 && running )
        found = i;
    }
    if ( found > 0 ) {
      System.err.println(
        "Promoting " + found + " to the top of the list" );
      todo.promote( found );
    }
    return Goal.States.PASSED;
  }
}
```

In the above plan, if an element named "abort" is found, it is promoted to the top
of the group. If it is not found, then the first element in the group which is wanting
to progress (i.e. for which `isRunning()` returns `true`) is promoted to the top
of the group. As we shall see in the last example in this chapter, this means that if
an instance object is added to a ToDo group in which all instance objects are
blocked, the added object will be promoted. The reason for this is that when the
executor adds an element to a ToDo group, the element is provided with an initial
state of `STOPPED + true`.

If required, more complex manipulation of the ToDo group can be performed.
In this regard, a meta-goal has access, via the data context, to the following
information:

Table 4.1 The data elements available to a meta-goal

Name	Type	Description
"added"	Vector	The Goal.Instance objects (if any) that were added to the ToDo group in its last execution cycle.
"removed"	Vector	The Goal.Instance objects (if any) were removed from the ToDo group in its last execution cycle.
"top"	Goal.Instance	The first Goal.Instance object in the ToDo group.
Goal.PERFORMER	Performer	The performer of the meta-goal.
Performer.TODOGROUP	ToDoGroup	The ToDo group associated with the meta-goal.

When a ToDo group contains multiple active intentions, a meta-goal can be used to progress the intentions in an appropriate manner. In this regard, goals are provided in com.intendico.gorite.addon to support round-robin progression of intentions (TodoGroupRoundRobin and TodoGroupParallel) and the progression of the first non-blocked intention (TodoGroupSkipBlocked). The difference between the two round robin goals is that Todo GroupParallel skips blocked intentions, whereas TodoGroupRoundRobin does not. If no meta-goal is specified, the default behaviour is that when the top element completes, the next element becomes the top element.

Another use of meta-goals is to determine which intention should be pursued when the currently executing intention terminates. As an example, consider the use of job lists in the scheduling of manufacturing operations. In this approach, jobs are allocated to individual machines. When a job completes on a particular machine, the next job to be performed is then selected from the list of available jobs for that machine using a suitable heuristic. Such a strategy is readily modelled in GORITE with a ToDo group acting as the job list and a meta-goal performing the selection process.

Note that meta-level reasoning in the above situations applies to the intentions that the performer is currently pursuing or is scheduled to pursue. This is in contrast to the BDI model, where meta-level reasoning applies to plans in the applicable set that can be employed to achieve a specific goal[3]. In our experience, reasoning about an agent's current intentions is much more common than reasoning about the plans in the applicable set.

Goal achievement through ToDo group execution can be seamlessly integrated with an intention being pursued through a `performGoal()` invocation, as we shall see in Chapter 5. Alternatively, ToDo group execution can itself drive intention realization, particularly in reactive environments. This is explored further in Section 4.3.

4.2 Action

In the BDI model, it is assumed that deliberation will result ultimately in action being performed in pursuit of the current intention. As indicated earlier, the BDI model makes no distinction in terms of strategic, tactical or operational reasoning – the primary focus of all reasoning is to determine what intention should be pursued and not how a particular intention should be progressed. The understanding is that the progression is detailed in pre-specified plans, although some level of dynamic customisation is achievable through the use of meta-level reasoning and plan context. However, at the operational level, progressing action invariably requires an understanding of the current situation in terms of the state of other intentions that are being pursued. For example, in the meter box example of the previous chapter, when both jigs are being used for assembly, the progress of the two assembly intentions are intertwined because of the involvement of shared resources, especially the rotating table. GORITE provides an elegant solution to such problems through the use of ToDo groups, which enable a performer to reason about how to progress its currently active intentions. At the same time, it does not preclude the performer from reasoning about which intentions should be pursued in order to achieve new goals.

As we have seen in the previous chapters, simple actions, such as greetings, can be modelled as goal instances having task semantics. These actions are simple in the sense that the action is completed in a single invocation of `execute()`, during which time the executor is blocked from pursuing other goals. However, if the action takes time, it is highly desirable to organise the execution so that the performer initiates the action, monitors its progress and notifies the executor of its completion. This way, the executor is then free to progress other goals while the action is in progress. What this means in practice is that the notion of an action lifecycle needs to be supported – an action, like a good story, will always have a beginning, a middle and an end. In keeping with its origins in Business Process Modelling, actions in GORITE are additionally viewed as transformational – that is an action operates on a specific set of inputs and produces a specific set of

---

[3] In GORITE, meta-level reasoning about the applicable set is supported through PlanChoice goals, as was explained in Section 3.1.

outputs. In this regard, inputs act as pre-conditions for the action to occur. To distinguish between the two action types, we refer to simple actions as *task goals* and to actions that take time as *action goals*.

From a design perspective, the distinguishing feature of GORITE's goal classes is the execution semantics, as was illustrated in Tables 3.1 and 3.3. In this regard, action goals will exhibit the same execution semantics as task goals with behaviour being defined by an execute() method. However, the execute() method for an action goal will have a different prototype to that used by a task goal in order to capture the nuances described above. So rather than

```
Goal.States execute( Data d )
```

the prototype for the execute() method for an action goal is

```
Goal.States
    execute(boolean reentry, Data.Element[] ins,
    Data.Element[] outs)
```

Note that an action goal is still an instance of the Goal class, but for reasons that will be explained below, actions are created using the Action factory class. Therefore, to avoid potential confusion, we will use *action* to refer to an instance of the Action factory class and *action goal* to refer to a goal created by an action factory.

In the prototype above, reentry is a flag set by the executor – it is false on the first invocation of execute() and true for all subsequent invocations. It is typically used to initiate the action goal behaviour on the first invocation and to monitor progress on subsequent invocations. As such, it supports a simple lifecycle model for the action goal. The inputs and outputs for the goal are provided by ins and outs respectively. The specification of the data elements that appear in the two arrays occurs when the action goal is created, using the action factory's create() method:

```
Goal create(String[] ins, String[] outs)
```

The elements in both arrays must all be members of the data context. Unlike a task goal, an action goal does not have automatic access to the data context through its execute() method. If access is required, it can be achieved by adding a data element to the data context that is a reference to the data context. Note that at action goal creation, the names of the elements are specified (as an array of type String), whereas at action goal execution, references to the actual data elements are provided by the executor (as an array of type Data.Element).

In terms of pre-conditions, an action goal's execute() method will only be invoked if all inputs contain data and are marked as being ready. In contrast, the execute() method for a task goal is invoked as soon as the goal is encountered in the executor's traversal of its enclosing process model. In order to realise this behaviour, action goals are created using the `Action` class, which is a factory for goal creation. When the value of an element is set using `Data.Element.set()`, the element is automatically marked as being ready. `set()` is invoked by both `Data.setValue()` and `Data.replace Value()`.

In Chapter 3, we presented the process models for the making of meter boxes as the behaviours of a single performer, namely the cell. In Chapter 5, we will see how those process models can be recomposed as team behaviours. In particular, the actual tasks required for assembly will be modelled as action goals performed by machines on behalf of the cell. We will now use one of those tasks, namely joining, to illustrate how action goals can be deployed in GORITE.

4.2.1 Version 1: Polling

In our first example, the executor will poll the progress of an activity. We begin by assuming that an instance of a suitably defined `Action` factory class called `join` has been created. `join` is then used to create an action goal that is added to the performer's default capability. This occurs inside the performer's constructor:

```
public Hirata( String name ) {
  super( name );

  // Create and add goals. Create takes as its arguments the
  // names of the elements in the data context that are to be
  // used as the inputs and outputs for the goal execution.
  // Execution will not proceed until all inputs have been
  // assigned values.
  addGoal( join.create(
    new String[] { Cell.JIG, Cell.ORDER }, null) );
}
```

`JIG` and `ORDER` are the names of data elements that will be added to the data context for the goal execution prior to the invocation of `performGoal()`. No data elements are used as outputs, so the second parameter for `create()` is set to `null`. The factory class is defined below – we specify the `execute()` method that goals that are created by the factory will employ.

```
Action join = new Action( Cell.JOIN_AB ) {
  public Goal.States execute(
    boolean reentry,Data.Element[] ins,Data.Element[] outs) {
    //construct a part identifier (count/jig)
    int j = ((Integer) ins[0].get()).intValue();
    int c = ((Order) ins[1].get()).count;
    String p = ""+c+"/"+j;
    //first invocation of execute?
    if ( !reentry ) {
        start( Cell.JOIN_AB, p );
        return Goal.States.STOPPED;
    }
```

```
    if ( busy( Cell.JOIN_AB, p ) )
        return Goal.States.STOPPED;
    return Goal.States.PASSED;
    }
};
```

In this case, `join` is an instance variable of the enclosing class (`Hirata`). Note that the data context for goal execution is **not** provided as a parameter to the `execute()` method and only the sub-set of its elements provided in `ins` and `outs` are accessible. When the data context is accessible (as in the `execute (Data d)` method for a goal instance), element values are retrieved via the element name using the `Data.getValue()` method. With an action goal, references to the actual elements are provided and element values are retrieved using the `Data.Element.get()` method.

In the above example, the re-entry flag supports a simple life-cycle model for actions, consisting of three phases:

1. Action initiation
2. Action realization and
3. Action completion

In this example, the behaviours associated with the respective phases are very simple. Action initiation is provided by the `start()` method:

```
public void start( String action, String part ) {
  System.err.println(
    "Hirata:\t started "+action+" on "+part+" at "+timestamp() );
  // t0 is an instance variable in the enclosing class
  t0 = System.currentTimeMillis();
}
```

`timestamp()` returns the time as a string formatted as *hh:mm:ss* :

```
String timestamp() {
  long now = System.currentTimeMillis();
  return String.format( "%tT", now );
}
```

Action realization is achieved by waiting for 5 seconds. On action completion, a message is printed indicating the time at which the action completed.

```
public boolean busy( String action, String part ) {
  long t = System.currentTimeMillis();
  if ( t-t0 < 5000 )
    return true;
  System.err.println( "Hirata:\t finished "+action+
    " on "+part+" at "+timestamp() );
  return false;
}
```

The output that is produced will be similar to the following:

```
Hirata:      started join AB on 0/0 at 14:53:01
Hirata:      finished join AB on 0/0 at 14:53:06
```

Note that as with goal instances, the executor associated with the action's performer will invoke the action's execute() method regularly through its lifecycle and the method will return appropriate values to indicate to the executor the status of the execution. On completion, PASSED is returned; all other calls will return STOPPED. In this case, STOPPED indicates that the goal execution has not completed, but is expecting re-entry as soon as possible; that is, it is expecting to be polled.

4.2.2 Version 2: Blocking

The alternative to polling is for the action to block. In this situation, BLOCKED is returned to indicate that the goal execution has not completed, and that it cannot progress until a blocking condition has been removed. When the action believes that the blocking condition may no longer apply, the executor is notified. The goal instance execution is then re-entered and the execute() method either returns PASSED if the action has in fact completed or it re-blocks and returns BLOCKED otherwise[4].

To illustrate blocking, we will modify the previous example to use a blocking delay. In our earlier examples, we have used TimeTrigger.pending() to model delays. In this example, we will reproduce that functionality by spawning a thread and then initiating a sleep on the thread for the required amount of time[5]. The reason for doing things this way, as opposed to using the TimeTrigger class, is that blocking is not restricted to time delays. For example, in the actual joining operation employed in (Jarvis et al., 2006), action completion was indicated by a bit in a status word of the PLC going high. In such instances, a separate thread would be spawned on the first invocation of execute() to initiate and monitor the action. When it completes, the executor will then be notified.

---

[4] While the need to confirm that the action has completed is good software engineering practice, it is in fact necessary in GORITE as under certain circumstances, the executor will need to re-enter the goal instance execution when it is blocked.

[5] In contrast, TimeTrigger achieves its delay by scheduling an event to occur on an instance of java.util.Timer. When the event occurs, it issues notifications to registered observers.

We begin our example by defining an `Alarm` class:

```
class Alarm extends java.util.Observable {
  int delay;
  public Alarm( int d ) {
    delay = d;
    Thread t = new Thread( new Runnable() {
      public void run(){
        try {
          Thread.sleep(delay*1000);
          Alarm.this.setChanged();
          Alarm.this.notifyObservers();
        } catch(Exception e) {}
      }
    });
    t.start();
  }
}
```

All that an alarm instance does is to create and start a new thread. When the thread runs, it immediately sleeps for the specified number of seconds. On waking up, it notifies any observers that have registered interest in the change by invoking the `setChanged()` and `notifyObservers()` methods provided by `java.util.Observable`. If you are unfamiliar with the observable/observer pattern or thread programming, refer to a comprehensive Java text, such as (Flanagan, 2005).

To use an alarm, we first need to provide the action with access to the data context so that the executor can be notified when the action is no longer blocked. As the action will have access to the data context, there is no need to specify input elements explicitly and as before, no output elements are used. The join action is then created as follows:

```
public Hirata( String name ) {
  super( name );

  // Create and add goals. Create takes as its arguments the
  // names of the elements in the data context that are to be
  // used as the inputs and outputs for the goal execution.
  // Execution will not proceed until all inputs have been
  // assigned values. The data context (DATA) was added as an
  // element of the data context in the goal initiation method in
  // order to give the goal execution access to its rendezvous
  // flag.
  addGoal( join.create(
    new String[] { Cell.DATA }, null) );
}
```

The observer that will be notified when the blocking condition changes is created automatically when the data context's setTrigger() method is invoked. This method takes an observable (in this case an Alarm instance) as its argument and creates an observer of that observable. When the blocking condition on the observable is removed, the observer's update() method will be invoked and the data context's rendezvous flag will be set. The executor will then know that the blocking condition has been removed and the goal's execute() method will be invoked. The definition of the action class is as follows:

```
Action join = new Action(Cell.JOIN_AB) {

  public Goal.States execute(
    boolean reentry, Data.Element[] ins, Data.Element[] outs) {
    // extract the data context.
    Data d = (Data) ins[0].get();

    //construct a part identifier (count/jig)
    int j = ((Integer) d.getValue(Cell.JIG)).intValue();
    int c = ((Order) d.getValue(Cell.ORDER)).count;
    String p = ""+c+"/"+j;

    // first invocation of execute?
    if ( !reentry ) {
      System.err.println(
        "Hirata:\t started joining "+p+" at "+timestamp() );
      // setTrigger() creates and binds an observer to an
      // observable (the alarm in this case) and then returns.
      d.setTrigger( new Alarm( 5 ) );
      return Goal.States.BLOCKED;
    }
    // When the alarm goes off, the update() method of the
    // observer that was created by setTrigger() is invoked. This
    // will result in the data context's rendezvous flag being
    // raised i.e. set to false. The executor will then invoke
    // execute(). However, the executor may need to invoke
    // execute() while the action is still blocked, so the
    // rendezvous flag needs to be checked.
    if ( !d.flag ) {
      System.err.println(
        "Hirata:\t finished joining "+p+" at "+timestamp() );
      return Goal.States.PASSED;
    }
    return Goal.States.BLOCKED;
  }
};
```

Note that while the data context was needed to access setTrigger(), access to the re-entry flag is also required in order to manage the action lifecycle. If the joining behaviour was to be modelled using an execute(Data) method, lifecycle management support would then become a developer responsibility.

4.3 Perception

In addition to providing modelling support for actions, GORITE also provides support for perception through the com.intendico.gorite.addon. Perceptor class. When using this infrastructure, observation of the environment or internal reflection by a performer results in the creation of an object, called a percept. Each percept type is handled by a separate Perceptor instance which specifies

1. the performer responsible for the handling of the percept[6],
2. the goal that will be initiated to handle the percept and
3. the ToDo group that will handle goal execution.

The data context provides the linkage between observation and goal execution; the perceptor adds the percept to a nominated data context as an element called "percept".

The overall process involves four steps and will be illustrated by means of a meter box assembly example:

   1. *Perceptor construction.*
       The general prototype for perceptor construction is

       Perceptor(Performer *p*, String *goal*, String *todo*)

       Variants are available, but in all cases, the performer must be specified. The default values for the goal and ToDo group are "handle percept" and "percepts" respectively. For example, to bind a perceptor to the current performer, the MAKE_BATCH goal and the default perception ToDo group, the following statement could be employed:

       ```
       // A perceptor for make meterbox requests.
       Perceptor request = new Perceptor( this, MAKE_BATCH );
       ```

   2. *Percept construction*
       In this step, an application specific object is created to reflect the particular observation that has taken place. We call this object a *percept*. For example, suppose that the cell performer from Chapter 3 was to monitor a TCP/IP socket for incoming job requests. An order object could then be created in response to an external event, rather than being generated internally to the application, as in Chapter 3.

---

[6] A perceptor can be created outside of the performer to which it is bound, in which case it will need to reference the performer. If it is to be bound to its enclosing performer, then this is used.

3. *Percept processing*

The processing of a percept is triggered by invocation of the perceptor's `perceive()` method. The general prototype for this method is the following:

`perceive(Object percept, Data data, String goal)`

If a goal is specified, then it is used in lieu of the goal that was specified when the perceptor was created. The percept is available to the goal's `execute()` method through the data context; it is bound to a data element named "percept". If a data context is not provided, one is created. For example, the statement

```
request.perceive( new Order( 10 ) );
```

will initiate processing for the percept (`new Order(10)`) using the perceptor's percept handling goal (`MAKE_BATCH`, in this case). No data context is specified, so a data context is created. No ToDo group was specified in the construction of the perceptor, so the `perceive()` method will add `MAKE_BATCH` to the default perception ToDo group ("percepts") for subsequent processing.

The percept processing code for our simplified example is shown below. Note that an external communications link is not monitored, as one would do in reality.

```
Thread t = new Thread ( new Runnable() {
  public void run() {
    try {
      Thread.sleep(2000);
      System.err.println(
       "cell: received request to make 10 meterboxes at "+
      timeStamp() );
      request.perceive( new Order( 10 ) );
    }
    catch( Exception e ) {}
  }
});
t.start();
```

4. *Goal creation*

The goal for handling the percept needs to be defined and created. The percept itself is made accessible to the goal execution through an element called "percept" that the perceptor will add to the data context when its `perceive()` method is invoked. If the default handling goal is used ("handle percept") be aware that it is a BDI goal and that the developer still needs to provide a goal (or goals) with that name.

Continuing with our meter boxes theme, we could reuse the make meter boxes goals from Chapter 3. Instead, we choose to abstract the making of meter boxes to a (very brief) delay and introduce a goal to retrieve the percept from the data context:

```
addGoal( new Plan( MAKE_BATCH, new Goal [] {
  new Goal ( PROCESS_PERCEPT ) {
    public States execute( Data d ) {
      Order o = (Order) d.getValue( PERCEPT );
          d.setValue( ORDER, o );
          return States.PASSED;
    }
  },
  new Goal ( MAKE_METERBOXES0 ) {
    public States execute( Data d ) {
      int n = ((Order) d.getValue( ORDER )).required;
      if ( TimeTrigger.isPending( d, "deadline", n*1000 ) )
          return Goal.States.BLOCKED;
      System.err.println(
          "cell: made "+n+" meterboxes at "+timeStamp() );
      return States.PASSED;
    }
  }
));
```

In Chapter 5, we will present an example of percept processing and its integration with process model execution realised through a performGoal() invocation.

GORITE also supports a pure percept processing execution model appropriate for event-driven systems. In discussing the four steps involved in percept triggered goal processing, we have already provided the basis of an event-driven batch manufacturing example. All that is missing from the example is the activation of the executor, as in an event-driven system we would not use performGoal() to initiate goal execution. In this example, the performer will automatically attach to the default executor, but the executor needs to be explicitly started. If we assume that the percept triggered goal processing behaviour discussed above is encapsulated in a performer of type Cell, then the application can be initiated with the following code:

```
public class Main {
  public static void main(String[] args) {

    // Start GORITE model execution
    new Thread( Executor.getDefaultExecutor(),"GORITE" ).start();

    // Event processing will be triggered from a separate thread
    // initiated in the cell's constructor.
    Cell c = new Cell( "Cambridge" );
  }
}
```

Goal execution is then initiated when goals are added to an appropriate ToDo group by the perceptor in the cell instance.

In the preceding example, percepts arise through observation of the external environment. Alternatively, percepts can correspond to changes in the beliefs of a performer, in which case percept processing is triggered by a process of internal reflection. GORITE provides the `Reflector` class for the modelling of such situations. We will defer a discussion of reflection until Chapter 6. For now, we will continue with our event processing theme and add a second perceptor.

In our next example, we retain the basic processing model (making meter boxes) of the previous example, but we now introduce the ability to stop processing immediately an emergency situation arises. This will require the promotion of the emergency stop/abort goal to the top of the default percept processing ToDo group ("percepts"), as described in Section 4.1.

In terms of process modelling, we need to add two goals to the cell performer's default capability that

1. abort processing and
2. manipulate the ToDo group to promote the abort goal when it is added to the ToDo group.

The goal to abort processing is defined as follows:

```
addGoal( new Plan( ABORT ) {
    // This goal tells the executor to relinquish its thread as
    // soon as possible. The goal also returns STOPPED so as to
    // "force" focus change, which in this case leads to the
    // execution finishing without pursuing any other goal
    // whatsoever.
    public States execute(Data d) {
        System.err.println("cell: abort");
        Executor.getDefaultExecutor().stop( true );
        return States.STOPPED;
    }
});
```

All that the abort goal needs to do is to notify the executor that execution is to stop (by invoking the executor's `stop()` method) and to ensure that the executor notices (by returning `STOPPED`).

The meta-goal to manipulate the ToDo group was presented in Section 4.1 and is not reproduced here. As indicated in Section 4.1, a meta-goal, if present, is executed at the beginning of every time slice. However, for it to be executed, the meta-goal must be bound to the ToDo group using the performer's `addTodoGroup()` method:

```
addTodoGroup( "percepts", "percepts meta goal" );
```

The above statement is added to the cell's constructor.

All that remains to be done is to ensure that the abort processing goal is added to the "percepts" ToDo group when an emergency stop is initiated. For this to occur, a perceptor must first be created and percept processing initiated. The perceptor is created as an instance variable of the cell performer:

```
Perceptor abort = new Perceptor( this, ABORT );
```

As in the previous example, percept processing is initiated from a separate thread. In this case, percept processing to initiate both MAKE_BATCH and ABORT goals is required. Also to better illustrate goal promotion, an additional order percept is created. The resulting code is shown below:

```
Thread t = new Thread ( new Runnable() {
  public void run() {
    try {
        Thread.sleep( 2000 );
        System.err.println(
          "cell: received request to make 10 meterboxes at "
          +timeStamp() );
        request.perceive( new Order( 10 ) );
        Thread.sleep( 1000 );
        System.err.println(
          "cell: received request to make 10 meterboxes at "
          +timeStamp() );
        request.perceive( new Order( 10 ) );
        Thread.sleep( 1000 );
        System.err.println(
          "cell: emergency stop at "
          +timeStamp() );
        // the percept is not used
        abort.perceive( null );
    }
    catch( Exception e ) {}
  }
});
t.start();
```

Note that the MAKE_BATCH goals that are initiated by the processing of the order percepts need to know the order size – this information is provided in the order percept and is copied from a data element named "percepts" to a data element called "order" by the PROCESS_PERCEPT sub-goal. In the case of the ABORT goal, no additional information is required, so the percept is set to null.

When the application is run, output similar to the following will be produced:

```
cell: received request to make 10 meterboxes at 13:55:17
reviewing percepts
[0] Goal make meterboxes0 STOPPED running
reviewing percepts
[0] Goal make meterboxes0 BLOCKED monitored
cell: received request to make 10 meterboxes at 13:55:19
reviewing percepts
[0] Goal make meterboxes0 BLOCKED monitored
[1] Goal make meterboxes0 STOPPED running
Promoting 1
reviewing percepts
[0] Goal make meterboxes0 BLOCKED monitored
[1] Goal make meterboxes0 BLOCKED monitored
cell: emergency stop at 13:55:20
reviewing percepts
[0] Goal make meterboxes0 BLOCKED monitored
[1] Goal make meterboxes0 BLOCKED monitored
[2] Goal abort STOPPED running
Promoting 2
cell: abort
```

When the application runs, the first order is received and its processing begins. The MAKE_METERBOXES0 goal returns BLOCKED as it has initiated a delay using TimeTrigger.isPending(). When the second order is received, its goal is added to the ToDo group by the executor with an initial state of STOPPED + running. It is then promoted, as it is running and the first goal is not. Finally, when the abort goal is added to the ToDo group, it is promoted and execution is stopped. Note that in this example, the meta-goal would promote the abort goal by virtue of its status when it is added to the ToDo group, as the other two goals are blocked – the matching on the goal name is in fact redundant. However, in general, this will not be the case.

Chapter 5
Team Modelling

In GORITE, no distinction is made between an agent and a team of agents in the sense that both have beliefs, desires and intentions. This modelling stance may seem unusual at first glance as we intuitively view team behaviour from an individual perspective, with team behaviour emerging from the contributions of individuals. However, if we abstract team member behaviour into roles, then team behaviour can be viewed as arising from a predefined interaction of roles, with individuals filling roles as in a play or a football side. Realization of team behaviour then requires a binding to be made between roles and individuals, but that process is separate from the definition of both the team behaviour and the individual behaviours.

As an example, consider a string quartet[1]. The string quartet team will have repertoire (team behaviours) and members who play roles (violin, violin, viola, cello). A particular piece (team behaviour) is defined in terms of the coordinated interaction of these four roles. Note that the number of members is not specified and can be greater than four. Also there is no correspondence assumed between roles and members – for example members could play multiple instruments. Team members can themselves be teams. For example, an orchestra will have a string section which in turn will have sub sections of violins, cellos, violas and double basses.

While string quartets and orchestras provide intuitive examples of team structure and behaviour, when one delves into the finer details, subtleties arise. For example, an orchestra will have one or more conductors. Since conductors are substitutable, should they be modelled as a role within an orchestra team, with the team behaviours (the musical score) residing with the orchestra team? However, conductors provide their own interpretation of a score, so one could argue that from the perspective of music delivery, the orchestra and the conductor are synonomous.

From a modeling perspective, the problem is that there are distinct team behaviours (scores) that are then interpreted/customized by the conductor. This is a recurring theme when human organizational systems are modelled. For example, a company commander will execute specific military doctrine (such as an attack on an enemy platoon), but different commanders will interpret the doctrine

---

[1] We are not suggesting that music for string quartets or orchestras is actually composed using a top down approach, but rather that it could be. For example, if one were to compose for a robotic string quartet, the top down approach would be a feasible strategy.

D. Jarvis et al.: Multiagent Systems and Applications, ISRL 46, pp. 77–124.
DOI: 10.1007/978-3-642-33320-0_5   © Springer-Verlag Berlin Heidelberg 2013

differently. One modelling approach is to ignore the problem and to treat the role and team as synonymous. Alternatively, one could model the role and the team separately, but then the interpretation of the team behaviours by the role would need to be modelled. In this book, our primary concern is with engineered systems (sensor networks and manufacturing cells) where there is little benefit to be gained by allowing team behaviours to be interpreted by the role tasked with their delivery. Consequently, we employ the first approach.

A further subtlety with team modelling is that it is useful to distinguish between the members of a team and the members that are actually involved in the realization of a behaviour. As was indicated above, a string quartet may have more than four members, but under normal circumstances, any particular performance will involve four members selected from the team. We call the team that is formed to perform a particular team behaviour a *task team*. Depending on the particular application, task teams can persist over multiple tasks (as in military mission management) or they can be reformed for each specific task (as in the fulfillment of a manufacturing order). Furthermore, task team membership may change while a particular behaviour is being performed – for example, when machines break down and as we shall see later in this chapter, sensors fail.

5.1 The Modelling Classes

Structurally, team members are of type `Performer`. The `Team` class is a subclass of the `Performer` class, as illustrated in Figure 5.1, so a team can be a team member of a team, including its own.

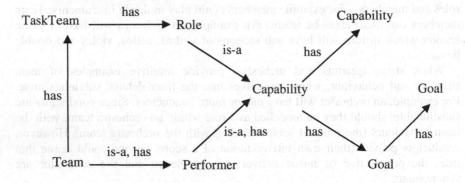

Fig. 5.1 GORITE team modelling classes (Rönnquist, 2012). The methods for the key team modelling classes are summarized in Appendix B.

A team structure is created by first creating a team instance and then creating and adding performers to the team, as shown below:

```
// Create a sensor network with 4 sensors and 1 supervisory node.

// Create the supervisory node.
Supervisor supervisor = new Supervisor( "supervisor" );

// Create and add sensors to the network. The sensors are able to
// sense temperature and humidity but are configured to sense
// only one property. The third parameter is the mode:
// 0 = Sensor.Controller.RELIABLE,
// 1 = Sensor.Controller.UNRELIABLE.
supervisor.addPerformers( new Performer[] {
  new Sensor( "sensor1", Sensor.TEMPERATURE, 0 ),
  new Sensor( "sensor2", Sensor.HUMIDITY, 0 ),
  new Sensor( "sensor3", Sensor.TEMPERATURE, 0 ),
  new Sensor( "sensor4", Sensor.HUMIDITY, 0 ),
});
```

In the above example, Supervisor extends Team and Sensor extends
Performer.

Actual team behaviour is realised by deploying organisational sub-structures
known as task teams. A task team is defined in terms of the roles that are required
in order to achieve its associated team behaviour, which will be defined as a
process model. In GORITE, a role is defined as set of related goals. Continuing
with our sensor network example, its sensing behaviour will be defined in terms of
the following TEMPERATURE_SOURCE and HUMIDITY_SOURCE roles[2]:

Table 5.1 Roles required for the supervisor team to be able to perform sensing

Role Name	Goal Name
TEMPERATURE_SOURCE	READ_TEMPERATURE
HUMIDITY_SOURCE	READ_HUMIDITY

As indicated earlier, a role can (and will normally) consist of multiple goals.
For example, in the assembly case study described in (Jarvis et al, 2006), the cell
team required the following roles in order to assemble a meter box (AB) from its
components (A and B):

---

[2] Both role instances and goal instances will normally be referred to by their name and not
their type. In this regard we continue the practice of previous chapters –
TEMPERATURE_ SOURCE and HUMIDITY_SOURCE are String constants, but we
forgo inclusion of their definition for the sake of brevity. In this book, goal instances and
role instances are normally created inline as anonymous class instances that subclass
either a GORITE Goal type or the Role type respectively. As such, they do not have an
explicit type and are not assigned to variables. In the interests of simplicity of expression,
when referring to particular role or goal instances we will often use the term role and goal
rather than role name and goal name.

Table 5.2 Roles required for the cell team to be able to make meter boxes (Jarvis et al., 2006)

Role Name	Goal Name
MOVER	MOVE_TO_JOINER MOVE_TO_LOADER RESERVE_JIGS EMPLOY_JIG RELEASE_JIG
LOADER	LOAD_A LOAD_B UNLOAD_AB
JOINER	JOIN_AB

A task team for performing sensing can be defined in the Supervisor class as follows:

```
//The task team definition
public TaskTeam providers = new TaskTeam() {
    {
        // the parameters for the role constructor are the role name
        // and the goal names
        addRole( new Role( TEMPERATURE_SOURCE, new String [] {
          READ_TEMPERATURE } ) );
        addRole( new Role( HUMIDITY_SOURCE, new String [] {
          READ_HUMIDITY} ) );
    }
};
```

Here, providers is an instance variable of the Supervisor class, so it is therefore accessible to all local and anonymous classes defined within that class.

At this stage, we have only defined the roles that the task team will need in order to perform its task. Before an actual team behaviour can be performed, these roles must be allocated to members of the Supervisor team. This process is called task team establishment. A task team represents a virtual structural unit within a team which is used to achieve particular team goals. As such, it is local to the team and does not need to be named in the same way as a performer or a goal. Nonetheless, it is beneficial from a modeling perspective to be able to label task teams within a team. For example, the intuitive way to model a manufacturing cell is as a hierarchy consisting of a cell (the team) and the machines comprising the cell (the team members). Now it may be that the cell (unlike the example used in this book) can make a range of different parts and furthermore, those parts may require different subsets of machines. Clearly, there is additional structure within the team, but that structure is fluid, not rigid and is dependent on the job mix that is presented to the cell. Labeled task teams provide a convenient way to model such goal-directed sub-structure. In this regard, GORITE supports both labeled and unlabeled task teams. We will refer to the two variants as *registered* and *unregistered* task teams to emphasise both the structural aspect and its fluidity.

Central to both registered and unregistered task teams is the use of the Task Team.establish() method to fill the roles required by the task team. By default, task team establishment will result in each of the required roles being filled by a single performer, where "filled" means that the performer is able to achieve **all** the goals required by the role. Furthermore, in the default strategy, a given performer will only perform a single role. As we shall see later in this chapter, the default strategy can be readily modified. Also note that while performers may group related goals into capabilities, GORITE assumes **no** correspondence between roles and capabilities. When a role is being filled, all the capabilities (and sub-capabilities) of a particular performer are considered.

The final aspect of team modelling in GORITE is the specification of team behaviour. As indicated above, the behaviour to be performed by a particular task team is specified as a process model. This specification is independent of the performers allocated to the task team; this independence is achieved through the use of team goals. As mentioned in Chapter 3, a team goal will appear as a leaf node in a process model and its execution is delegated to the filler of the role. The filler will then provide one or more process models with the same name as the delegated goal for execution. These goals will form an applicable set which will be executed in the normal manner. Note that if there are multiple fillers for a role, goal execution is performed concurrently in the same manner as a repeat goal.

A team goal is created with the following constructor:

```
TeamGoal( String role name, String goal_name )
```

5.2 Developing Team Applications

Given that team programming is at such an early stage of development, it would be presumptious of us to present a design methodology. Indeed, design methodologies for traditional BDI systems is still an active area of research even after 25 years of deployment. Consequently what we will do instead is to provide a design outline for the examples in this chapter. This will serve two purposes – it will provide a reference point for the subsequent discussion of the examples and a starting point for the design of your own applications. We begin with a brief discussion of the overall structure.

The key artefacts for our team applications are the following classes:

1. A class that provides the application entry point, i.e. a main() method
2. One class for each team type
3. One class for each team member type

Each of these classes will be contained in its own source file. The application
entry point is responsible for creating and initialising teams and their members and
also for the initiation of team behaviour. All of these interactions are through
method calls.

As with the single performer case, goal execution is managed by an executor
object which makes the data context available to all team members involved in a
goal execution[3]. While the data context provides the primary mechanism for
interaction between team members, direct interaction between team members is also
allowed. However, as direct interaction is not formally part of the GORITE
execution model, no explicit support for such interaction is provided. If required, it
can be realised using Java library functionality, such as java.net, java.rmi or
the Observer/Observable classes of java.util. Alternatively, third party
software can be employed.

The overall application structure is summarised in Figure 5.2:

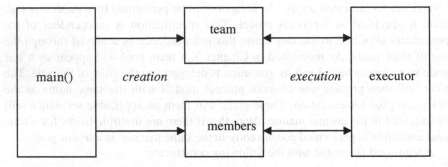

Fig. 5.2 The application structure employed for the examples in this chapter

In the examples in this chapter the responsibilities of the application classes are
as follows:

5.2.1 The Main Class

1. Creation of team instances.
2. Creation of team member instances.
3. Addition of team members to teams using Team.Add
    Performers().
4. Initiatiation of goal achievement by the team by invocation of the team's
    goal initiation method.

Note that 2 and 3 will normally be combined into a single statement.

---

[3] By default, an application has a single executor to which a performer is bound at
construction time. Additional executors can be created, in which case the binding between
performer and executor is explicit, using the executor's addPerformer() method.

5.2.2 *The Team Class*

1. Definition of the process models that the team is able to perform and
    their addition to the team's capability structure.

    Process model addition is realised using the team's `addGoal()`
    method; `addGoal()` invocations are normally placed in the team's
    constructor. In the absence of an explicit capability structure, goals are
    added to the team's default capability. The models to be added can either
    be defined inline, using anonymous classes or in separate files. As
    discussed in the previous chapter, a process model can incorporate
    instance methods that return process models/goals.

2. Definition of the task teams that the particular team will employ.

    Task teams will typically be defined as anonymous classes, which can
    then be assigned to an instance variable or bound to a name by using the
    `Team.setTaskTeam()` method. If the latter approach is used, then the
    invocation would normally be placed in the team's constructor. The task
    team definitions consist of a sequence of `TaskTeam.addRole()`
    methods, with the roles to be added typically being defined inline using
    anonymous classes.

3. Establishment of the task teams.

    Each process model that a team can perform will require a suitable
    task team to be established before the model can be achieved. Task team
    establishment requires that the task team's `establish()` method is
    invoked. This invocation can be either direct or indirect depending on
    whether the task team is unregistered or registered. If it is unregistered,
    then the task team's `establish()` method is invoked directly on the
    task team instance, typically in the team's goal initiation method.
    However, if the task team is registered (by using `Team.set`
    `TaskTeam()`), then the team's `deploy()` method is added to the
    definitions of the process models that will use the registered task team.
    The `deploy()` method takes a single argument, which is the name of
    the task team to be used and returns a goal that will result in the
    `establish()` method for the named task team being invoked.

    If the default role filling strategy (a distinct and singular role filling,
    with the performer able to achieve all of the goals required by the role) is
    not appropriate, then alternative strategies can be provided by overriding
    the `Role.canAct()` and `Role.canFill()` methods. These
    customisations, if required, are typically incorporated into the task team
    definition.

4. Provision of goal initiation methods.

    Goal execution will be managed by one or more goal initiation
    methods. A goal initiation method will

- Create the data context for the goal execution and create and initialise any data elements.
- For unregistered task teams, establish the task team via a call on its `establish()` method. Note that the data context is passed as a parameter to the `establish()` method; `establish()` adds the role fillings to the data context.
- Achieve the goal via a `performGoal()` invocation.

5.2.3 The Team Member Classes

As a team member can be a team, the generic functionality that a team member can provide is the same as for a team. However, in the examples in this book, team members are not teams and consequently functionality is restricted to the definition of the process models that the team member is able to perform and their addition to the team member's capability structure.

5.3 Hello World

The story so far …

> *Paul has landed on earth to greet the Earthlings. His efforts have had mixed success – landing in Antarctica and talking to the penguins was a lowlight. So how did Paul get here? Paul is a Martian and a Martian spacecraft is flown in a particular way. It requires a pilot to fly the spacecraft and a crew to look out for the destination. Once at the destination, the pilot stops flying, and the Martian who has been selected for the greeter role, greets the Earthlings. Paul is a greeter.*

This story suggests a design in which we have a `SpaceShip` team with a number of Martians as team members. The primary team behaviour will be a `VISIT_PLANET` goal which will be defined in terms of `PILOT`, `CREW` and `GREETER` roles. The definition for these roles will involve only a single goal per role, as shown below:

Table 5.3 The role definitions required for achieving the `VISIT_PLANET` goal

Role Name	Goal Name
PILOT	FLY
CREW	LOOKOUT
GREETER	GREET

In this example, the roles identified In Table 5.3 will be performed by individual performers. While one would expect the PILOT and GREETER roles to be performed by individual performers, the CREW role could be delegated to multiple performers. The LOOKOUT goal would then succeed when the first of the performers filling the CREW role observes the destination. GORITE provides for multiple ways of modelling this situation, including the preferred way of attributing to the role object a lookout plan that is wrapped in a control goal. A discussion of the finer points of the modeling of team behaviour is left to a future time – for now, the GREETER role is modelled by a single performer.

5.3.1 The Main Class

The application's main () method is as follows:

```
public class Main {
    public static void main(String [] args)throws Throwable {
        SpaceShip ship = new SpaceShip( "Enterprise" ) {{
            addPerformers( new Performer [] {
                new Martian( "Ralph" ),
                new Martian( "Dennis" ),
                new Martian( "Jacquie" ),
                new Martian( "Paul" )
            } );
        }};
        ship.visit("earth");
    }
}
```

It follows the pattern presented earlier in this chapter, namely that the application's entry point is responsible for

1. Creating the team
2. Creating the team members and adding them to the team and
3. Invoking the team's goal initiation method

5.3.2 The SpaceShip Team

We begin by considering the SpaceShip team. Before the team behaviour required for Paul's mission can be defined, a task team needs to be created. Although the task team will only be used for the achievement of the VISIT_PLANET goal, we will create it as a registered task team. As noted earlier, the distinction between a registered and unregistered task team is that a registered task team has a name that is known to the team and which can be used to access the task team instance. In this case, the name will be TRAVEL and the

binding between name and task team is achieved with the Team.set
TaskTeam() method:

```
setTaskTeam( TRAVEL, new TaskTeam() { {
    addRole( new Role( PILOT,    new String [] { FLY } ) );
    addRole( new Role( CREW,     new String [] { LOOKOUT } ) );
    addRole( new Role( GREETER, new String [] { GREET } ) );
} } );
```

This method is invoked from the team's constructor. Note that all that we have
done at this stage is to define and create an instance of the task team – no roles
have been filled.

Next, we need to define the mission behaviour (VISIT_PLANET) in terms of
the above roles.

We do this by using team goals, as shown below. Achievement of a team goal
is delegated to the task team member that has been allocated to the specified role.
In this example, establishment of the task team (TRAVEL) is achieved by the goal
returned by the team's Capability.deploy() method.

```
new SequenceGoal( VISIT_PLANET, new Goal [] {
    deploy( TRAVEL ),
    new ParallelGoal( FLY_TO_PLANET, new Goal [] {
        new TeamGoal( PILOT, FLY ),
        new ControlGoal( ARRIVED, new Goal [] {
            // look out for arrival
            new TeamGoal( CREW, LOOKOUT )
        } )
    } ),
    new TeamGoal( GREETER,  GREET )
} )
```

In the expression above, the top level sequence goal, VISIT_PLANET has two
sub-goals, FLY_TO_PLANET and GREET. These two goals are achieved in
sequence. Note that we could have replaced the sequence goal with a plan in
which neither context or precedence are overridden.

FLY_TO_PLANET has two sub-goals that are performed in parallel:

1. the flying of the spaceship by the pilot and
2. the monitoring of flight progress to determine arrival by the crew

As noted in Chapter 3, a parallel goal consists of two or more branches that are
executed concurrently. In this case there are two branches – flying the spaceship
and monitoring flight progress. In the absence of a control goal, a parallel goal

will succeed when all its branches succeed. This strategy doesn't work here, as we will be modelling FLY as a non-terminating activity so it will never succeed. Consequently, LOOKOUT is wrapped in a control goal. Thus when LOOKOUT succeeds, the control goal succeeds, all parallel branches (i.e. FLY) are cancelled, the enclosing parallel goal (FLY_TO_PLANET) succeeds and the mission progresses to the greeting phase.

In summary, the VISIT_PLANET goal is achieved as follows:

1. First the TRAVEL task team is established. The performer that fills the PILOT role is asked to achieve a FLY goal and in parallel, the performer that fills the CREW role is asked to achieve the LOOKOUT goal.
2. When the LOOKOUT goal succeeds, the ARRIVED goal succeeds, the FLY goal is cancelled, and the FLY_TO_PLANET goal succeeds.
3. The performer that fills the GREETER role is then asked to achieve the GREET goal.
4. When the GREET goal has been achieved, then the VISIT_PLANET goal has succeeded.

The VISIT_PLANET goal needs to be added to the team's default capability. This is achieved through an appropriate invocation of the addGoal() method in the team's constructor.

Finally, we need to provide a mechanism for initiating the VISIT_PLANET intention. In this case, we will provide an instance method visit() that prepares the data context for the intention and activates the intention through an invocation of performGoal():

```
public boolean visit(String p ) {
    // set the planet to be visited
    Data data = new Data().setValue( PLANET, p );
    return performGoal(
        new BDIGoal(VISIT_PLANET),"visit",data );
}
```

5.3.3 The Crew Members

The final step required for the completion of the application is the definition of the performer behaviours. Martians are multi-skilled and all team members are able to perform all roles that are required. Also, we admit the possibility that the roles that have been identified may expand in terms of the number of goals that they require. Therefore, rather than model the crew members as performers who have individual goals, we introduce a Martian class that subclasses Performer and has Piloting, Stargazing and Greeting capabilities that have the goals shown in Table 5.4:

Table 5.4 Martian capabilities

Capability	Goal Name
Piloting	FLY
Stargazing	LOOKOUT
Greeting	GREET

In this instance, there is a 1-1 mapping between the content of team roles and the content of capabilities. In general, this will not be the case.

The Greeting capability is defined as follows:

```
public class Greeting extends Capability {
  public Greeting() {
    addGoal( new Goal( SpaceShip.GREET ) {
      public Goal.States execute(Data d) {
        System.err.println(
          "Hello "+d.getValue( SpaceShip.PLANET ) );
        return Goal.States.PASSED;
      }
    } );
  }
}
```

The GREET goal is the same as in our first Hello World example.

The Greeting capability above is quite simple. In general a capability may include any number of goal hierarchies of any complexity as well as sub-capabilities of any complexity. Furthermore, as it is a bona-fide Java class, it may include any methods and members that might be needed in order to implement the desired function or functions. Capabilities may also be built dynamically, in which case they exist as particular unnamed competences rather than types.

The other two capabilities are marginally more interesting. First, we have the Piloting capability:

```
public class Piloting extends Capability {
  public Piloting() {
    addGoal( new Plan( SpaceShip.FLY ) {
      public Goal.States execute( Data d ) {
        System.err.printf(
          "Flying to %s\n",
          d.getValue( SpaceShip.PLANET )
        );
        return Goal.States.BLOCKED;
```

```
        }
     } );
   }
}
```

Note that the `execute()` method for `FLY` returns `BLOCKED`, which means that the goal will neither pass nor fail, which is what we require.

Finally we have the `Stargazing` capability:

```
public class SpaceGazing extends Capability {
  public SpaceGazing() {
    addGoal( new Plan( SpaceShip.LOOKOUT, new Goal [] {
      new Goal( "arrived" ) {
        public Goal.States execute(Data d) {
          // TimeTrigger is defined in com.intendico.gorite.addon
          // isPending() sets an alarm for 1000 msec in the
          // future.Until it goes off, BLOCKED is returned.
          if ( TimeTrigger.isPending( d, "deadline", 1000 ) )
            return Goal.States.BLOCKED;
          System.err.printf(
            "Arrived at %s\n",
            d.getValue( SpaceShip.PLANET )
          );
          return Goal.States.PASSED;
        }
      }
    } ) );
  }
}
```

As indicated earlier, Martian space crew are multi-skilled and all crew members possess all three capabilities. We model this by defining a `Martian` performer type:

```
public class Martian extends Performer {
  public Martian(String name) {
    super( name );
    addCapability( new Piloting() );
    addCapability( new SpaceGazing() );
    addCapability( new Greeting() );
  }
}
```

And that completes our example. When the application is run, output similar to the following will be produced:

```
Flying to earth
Flying to earth
```

```
Flying to earth
Arrived at earth
Hello earth
```

Note that the `execute()` method for the `FLY` plan is invoked more than once, even though it returns `BLOCKED`. The reason for this is that a performer's `execute()` method is invoked by the executor every time it is the performer's turn in the fair succession of progressing intentions and it has to keep claiming to be blocked (and resetting its trigger if one is set), until it finds that it can progress.

An implementation of the space travel example is provided with the GORITE distribution in the `tests` directory. It is essentially the same as the example presented here, except that the `VISIT_PLANET` intention is triggered (and the application terminated) through the use of percepts, rather than through a `performGoal()` invocation.

5.4 The Meter Box Cell

In this version of the meter box cell, we revisit our final meter box example of Chapter 3, namely using both jigs to make a batch of meter boxes. However, this time the meter box cell will be modelled as a team consisting of a pick and place robot (Fanuc), a screwdriver robot (Hirata) and a rotating table (Table) rather than as a single performer as in Chapter 3.

We begin this example by considering the goals that were required to be performed by the Cell performer of Chapter 3. Firstly, the goals are allocated to performers (Table 5.3) and then to roles (Table 5.4):

Table 5.3 Performer responsibilities

Goal	Performer
RESERVE_JIGS	Cell
MAKE_BATCH	Cell
MAKE_METERBOXES	Cell
MAKE_METERBOX	Cell
MOVE_TO_LOADER	Table
LOAD_A	Fanuc
LOAD_B	Fanuc
MOVE_TO_JOINER	Table
JOIN_AB	Hirata
UNLOAD_AB	Fanuc
RELEASE_JIGS	Cell

Table 5.4 Role allocations

Goal	Role
LOAD_A	Loader
LOAD_B	Loader
MOVE_TO_JOINER	Mover
JOIN_AB	Joiner
MOVE_TO_LOADER	Mover
UNLOAD_AB	Loader

As in (Jarvis et al., 2006), three roles are employed – Mover, Loader and Joiner and the mapping between performers and the roles that they can perform is 1-1, in contrast with the previous example, where the mapping was 1-many. One point of departure from Jarvis et al. is that RESERVE_JIGS and RELEASE_JIGS are not included in the MOVER role. Rather, we choose to model them as cell goals. This reflects a design choice that was not available for the earlier implementation, namely that GORITE allows the state of the jigs to be modelled as data context – jig states do not need to be bound to a particular performer/agent. Thus both the cell and the table can have their reasoning informed by the status of the jigs. In the case of the cell, this reasoning will relate to jig allocation, whereas for the table, it will relate to table movement.

5.4.1 The Main Class

The application's main() method is as follows:

```
public class Main {
  public static void main(String [] args)throws Throwable {
    Cell cell = new Cell( "cambridge" );
    cell.addPerformers(new Performer[] {
      new Fanuc( "fanuc" ),
      new Hirata( "hirata" ),
      new Table( "table" )
    });
    // Assemble 10 meterboxes
    cell.assemble(10);
  }
}
```

It follows the pattern presented earlier in this chapter, namely that the application's entry point is responsible for

1. Creating the team
2. Creating the team members and adding them to the team and
3. Invoking the team's goal initiation method

5.4.2 The Cell Team

We begin by creating an unregistered task team (assemblers) for making meter boxes:

```
// define the task team structure in terms of the roles required
public TaskTeam assemblers = new TaskTeam(){ {
    // The cell has a "mover" sub system, for moving
    // jigs holding parts between the stations of the cell.
    addRole( new Role( MOVER, new String [] {
        MOVE_TO_JOINER, MOVE_TO_LOADER }));

    // The cell has a "loader" sub system, for
    // loading and unloading the mover's carrier units.
    addRole( new Role(LOADER, new String[] {
        LOAD_A, LOAD_B, UNLOAD_AB }));

    // The cell has a "joiner" sub system, for
    // joining parts loaded into a jig.
    addRole( new Role(JOINER, new String[] {
        JOIN_AB }));
}};
```

In terms of behaviour, task team establishment will not be modelled as goal achievement as in the previous example. Rather, it will be performed procedurally before goal execution via the following statement in the goal initiation method:

```
// establish() establishes assemblers using the default strategy
// use the first performer that performs the role. It then adds
// the role fillers to the data context.
assemblers.establish( data );
```

Consequently, the process model structure from Chapter 3 can be retained, but with the actual assembly operations being specificied as team goals, rather than goals. In this regard, the top level goal instance method, makeBatch3(), which was defined as follows:

```
Goal makeBatch3() {
  return new SequenceGoal( MAKE_BATCH_3, new Goal[] {
    step( RESERVE_JIGS, 0 ),
    splitBatch(),
    step( RELEASE_JIGS, 0 ),
  });
}
```

now becomes:

```
Goal makeBatch() {
  return new SequenceGoal( MAKE_BATCH, new Goal[] {
    reserveJigs(),
    splitBatch(),
    releaseJigs()
  });
}
```

In this example, the assembly operations will be performed by task team members and not by the cell, as in Chapter 3. These operations are encapsulated in the makeMeterBox() method, so splitBatch() remains the same, as does makeMeterBoxes():

```
Goal splitBatch() {
  // The strategy that we are employing is to use all available
  // jigs (2) for assembly. Each jig will have its own branch in
  // a RepeatGoal. JIG is the control element.
  return new RepeatGoal( JIG, SPLIT_BATCH, new Goal [] {
    makeMeterboxes()
  });
}

Goal makeMeterboxes() {
  return new LoopGoal( MAKE_METERBOXES, new Goal [] {
    checkOrder(),
    makeMeterbox(),
    updateOrder(),
  });
}
```

The assembly operations now need to be delegated to cell team members, rather than being achieved directly by the cell, as in Chapter 3. The `makeMeterBox()` method is therefore realized using team goal instances rather than goal instances:

```
Goal makeMeterbox() {
  return
    new SequenceGoal( MAKE_METERBOX, new Goal[] {
        new TeamGoal( MOVER, MOVE_TO_LOADER ),
        new TeamGoal( LOADER, LOAD_A ),
        new TeamGoal( LOADER, LOAD_B ),
        new TeamGoal( MOVER, MOVE_TO_JOINER ),
        new TeamGoal( JOINER, JOIN_AB ),
        new TeamGoal( MOVER, MOVE_TO_LOADER ),
        new TeamGoal( LOADER, UNLOAD_AB )
      }
  );
}
```

As discussed earlier, the status of the jigs is required for reasoning related to both jig allocation (a cell responsibility) and table movement (a table responsibility). This status information is maintained in an element of the data context named JIGS, which is of type `Jigs`:

```
public class Jigs {
  public State[] states;
  public int allocated;

  public class State {
    public boolean inUse;
    public boolean moveRequested;
    public int atStation;

    public State( boolean flag1, boolean flag2, int i ) {
      inUse = flag1;
      moveRequested = flag2;
      atStation = i;
    }
  }

  public void reserve( int n ) {
    for ( int i = 0; i < n; i++ )
        states[i].inUse = true;
    allocated = n;
  }

  public void release( ) {
    allocated = 0;
  }
```

```
public Jigs() {
  states = new State[] {
    new State( false, false, Table.STATION0 ),
    new State( false, false, Table.STATION1 ),
  };
  allocated = 0;
}
}
```

The functionality provided by the `Jigs` class is minimal, but sufficent for our immediate purposes.

The RESERVE_JIGS goal reserves the appropriate number of jigs by invoking the `Jigs.reserve()` method. The goal behaviour is defined below:

```
public Goal reserveJigs() {
  return new Goal( RESERVE_JIGS ) {
    public Goal.States execute( Data d ) {
      int n = (Integer) d.getValue( JIGS_REQUESTED );
      Jigs j = (Jigs) d.getValue( JIGS );
      switch ( n ) {
        case 0:
        case 2:
          //use both jigs
          j.reserve( 2 );
          d.setValue( JIG, Table.JIG1 );
          d.setValue( JIG, Table.JIG0 );
          System.err.println("Cell:\t jig0 and jig1 reserved");
          break;
        case 1:
          //use one jig (JIG0)
          j.reserve( 1 );
          d.setValue( JIG, Table.JIG0 );
          System.err.println( "Cell:\t jig0 reserved" );
          break;
        default:
          System.err.println( "Cell:\t only 2 jigs available" );
          System.exit( 1 );
      }
      return Goal.States.PASSED;
    }
  };
}
```

The `jigs` object is accessed through the data context. The goal also needs to create and populate the multi-valued data element named JIG, which is used as the control value for the SPLIT_BATCH repeat goal. In this respect, note that multiple invocations of `Data.setValue()` produce multiple values for the named data element. To overwrite the topmost value for a named data element, use `Data.replaceValue()`.

The behaviour embodied in the RELEASE_JIGS goal is likewise straight forward:

```
public Goal releaseJigs() {
  return new Goal( RELEASE_JIGS ) {
    public States execute( Data d ) {
        Jigs j = (Jigs) d.getValue( JIGS );
        j.release();
        System.err.println( "Cell:\tjigs released");
        return States.PASSED;
    }
  };
}
```

In this example, the MAKE_BATCH goal will be stored in the cell team's default capability. This is achieved by the inclusion of the following statement in the team's constructor:

```
addGoal( makeBatch() );
```

Finally, a goal initiation method (assemble()) needs to be defined that performs the following tasks:

1. Create the data context
2. Establish the assemblers task team
3. Perform the MAKE_BATCH goal

The following method was used in this example:

```
boolean assemble(int n) {

  // set data context
  Data data = new Data();
  data.setValue( ORDER, new Order( n ) );
  data.setValue( JIGS_REQUESTED, 2 );
  data.setValue( JIGS, new Jigs() );

  // establish() establishes assemblers using the default
  // strategy - use the first performer that can performs the
  // role. It then adds the role fillers to the data context.
  assemblers.establish( data );

  // make boxes
  return performGoal(
    new BDIGoal( MAKE_BATCH ), "assemble", data );
}
```

In terms of the data context, firstly an element (ORDER) is needed to

1. track of the number of boxes that are being assembled,
2. track of the number of boxes that have been assembled and
3. specify the number of assembled boxes that are required.

A minimal class definition suitable for our purposes is provided below:

```
public class Order {
  public int wip;          // number being assembled
  public int count;        // number that have been assembled
  public int required;     // number required

  public Order(int r) {
    required = r;
    count = 0;
    wip = 0;
  }
}
```

Next, the RESERVE_JIGS goal needs to know how many jigs it should attempt to allocate to the current order – this information is provided by the JIGS_REQUESTED element. Finally, the JIGS element is provided to maintain the status of the jigs in order to manage both jig allocation and table movement.

5.4.3 Team Member Classes

Having defined the behaviours for the cell, the next issue is to define the behaviours for its team members. Recall that in our initial design, we chose to have a 1-1 mapping between the roles defined in Table 5.4 and the cell team members. That is, the rotating table is responsible for the achievement of all goals that constitute the MOVER role, the Fanuc robot for all goals in the LOADER role and the Hirata robot for all goals in the JOINER role[4]. These goals are all achieved directly by the team members concerned, and as they all correspond to physical tasks that take time, they will be modelled as action goals and not task goals.

We will begin by considering the performer for the Hirata robot. For simplicity, we will use polling of the performer by the executor to monitor action progress. The example provided in Chapter 4 provides the basic functionality for the performer. For completeness, we provide the entire class definition below:

---

[4] This design was chosen for simplicity of presentation – one could argue that the behaviour required to provide a MOVER role is better modelled as team behaviour. In that case, the mover team would consist of three team members – the table and the two jigs.

```
public class Hirata extends Performer {

    // If a performer's TODO_GROUP variable is set to a non-null
    // value,a ToDo group will be automatically created and used as
    // the default for goal execution. In this case, we are using
    // the ToDo group to sequentialise the tasks that the Hirata
    // performs. Whenever an assembly stream requires a joining
    // operation, the task is added to the to do group and is
    // executed on a first in,first out basis. No reasoning is
    // invoked on either task addition or task completion.
    private String TODO_GROUP = "Hirata Job List";

    // the start time for the current task
    private long t0;

    public boolean busy( String action, String part ) {
        long t = System.currentTimeMillis();
        if ( t-t0 < 5 )
            return true;
        System.err.println("Hirata:\tfinished "+action+" on "+part);
        return false;
    }


    public void start( String action, String part ) {
        System.err.println("Hirata:\t started "+action+" on "+part);
        t0 = System.currentTimeMillis();
    }

    Action join = new Action( Cell.JOIN_AB ) {
        public Goal.States execute(
            boolean reentry,Data.Element[] ins,Data.Element[] outs) {

            //construct a part identifier (count/jig)
            int j = ((Integer) ins[0].get()).intValue();
            int c = ((Order) ins[1].get()).count;
            String p = ""+c+"/"+j;

            //first invocation of execute?
            if ( !reentry ) {
                start( Cell.JOIN_AB, p );
                return Goal.States.STOPPED;
            }

            if ( busy( Cell.JOIN_AB, p ) )
                return Goal.States.STOPPED;
            return Goal.States.PASSED;
        }
    };

    public Hirata( String name ) {
        super( name );

        // Create and add goals. Create takes as its arguments the
        // names of the elements in the data context that are to be
        // used as the inputs and outputs for the goal execution.
        // Execution will not proceed until all inputs have been
        // assigned values.
        addGoal( join.create(
            new String[] { Cell.JIG, Cell.ORDER }, null) );
    }
}
```

Note that the performer is provided with a ToDo group, which will then be used to manage the execution of its goals. The purpose of the ToDo group is to sequentialise goal execution requests from the potentially multiple assembly streams.

The Fanuc class follows the same pattern as the Hirata class, except that where the Hirata performer can only perform one action (JOIN_AB), the Fanuc performer can perform three actions – LOAD_A, LOAD_B and UNLOAD_AB. Given the similarity of the behaviours, the code for the Fanuc performer is not reproduced.

As with the Fanuc and Hirata goals, the Table's MOVE_TO_LOADER and MOVE_TO_JOINER goals are modelled as actions and realised as time delays. However, the logic underpinning the table actions is more complicated for the following reasons:

1. The logic for table movement is different depending on whether one or two jigs are in use.
2. The table may already be in the correct position, in which case no action is required.
3. The table must not be moved if a joining or loading/unloading operation is in progress.

To address the third issue, we constrain table movement to occur only when both jigs have requested a move. Note that this coordination strategy does not require explicit resource locking. Also, it adds nothing of interest from a GORITE perspective and as the code is relatively lengthy, it is omitted. The complete application (together with all the applications presented in this book) are available from the GORITE web site.

One of the consequences of the table movement strategy that we have implemented is that the assembly of the last meter box can become a special case. Recall that a branch will terminate if it determines that the other branch is assembling the final meter box. The strategy of moving the table only when both branches have requested moves will then fail as the table will wait indefinitely for a second request. The solution that we have employed is to decrement the number of jigs allocated when a branch completes – table movement for the remaining branch is then handled by the single jig case. The modified code for the checkOrder() method is shown below:

```
EndGoal checkOrder() {
  return new EndGoal( CHECK_ORDER ){
    public Goal.States execute( Data d ) {
    // An order object specifies the number of boxes required
    // (required) and keeps track of the number of boxes that
    // have been assembled (count).
      Order o = (Order) d.getValue( ORDER );

    // has the goal been achieved?
```

```
    if ( o.count >= o.required ) {
      System.err.println( "Order filled" );
      return Goal.States.PASSED;
    }
      if ( o.count + o.wip >= o.required ) {
        System.err.println( "Order about to be filled" );
        Jigs jigs = (Jigs) d.getValue( JIGS );
        jigs.allocated--;
        return Goal.States.PASSED;
      }

      o.wip++;
      return Goal.States.FAILED;
    }
  };
}
```

5.5 Sensor Networks

In our discussion of the team modelling classes in Section 5.1, we used sensor networks as a motivating, but simple example. To recap, we chose to model a sensor network as a hierarchy of sensor nodes and supervisor nodes. The sensors are multi-purpose, being able to read both temperature and humidity. However, a particular sensor is configured to read either temperature or humidity. A supervisory node is responsible for a number of sensors – its team members. It will read the values of its members according to a predefined regime – in these examples, it take readings every 5 seconds from one temperature sensor and one humidity sensor.

Having sampled its sensors, the supervisory node will validate the readings and transmit data to the base station, again, according to a predefined regime. That aspect of the node's behaviour is not of interest in the following examples – our focus is on the organizational and behavioural issues associated with team reformation. In this regard, there are two drivers for team restructuring:

1. A sensor node may fail. In this case, we assume that the supervisory node has two strategies that it can employ. First, it can reset the sensor. If that does not work, then the task team is reformed with the remaining team members. If a task team can't be formed, that is, there is not a working temperature sensor and a working humidity sensor among the team members, then the supervisor's sensing intention fails.
2. A supervisor node may fail. We assume that failure only arises from loss of power. Power is monitored by the supervisor, so it is able to take remedial action when loss of power is imminent. This action will consist of the transfer of its team members to other supervisory nodes.

Note that implicit in the above discussion is a separation between device behaviour and device embodiment. We are dealing with physical devices that perform specific actions on their environment. For example, a sensor senses the current temperature or humidity. That functionality is provided by what we call its embodiment – that is, actual sensing hardware. Behaviour on the other hand, refers to how device functionality is achieved. In the case of sensors, the behaviours correspond more or less directly to embodiment functionality. However, as we shall see in the following examples, supervisor behaviours can be composed from both sensor behaviours and supervisor embodiment functionality, yielding much richer behaviours than in the case of sensors. This separation of embodiment from behaviour has been employed successfully in both the manufacturing sector (Jarvis et al., 2006) and the defence sector (Jarvis et al., 2005). A consequence of the approach is that the behaviour parts for a team can be deployed separately from the embodiment parts and furthermore, the behaviour parts do not necessarily need to be physically distributed, leading to deployments like Figure 5.2:

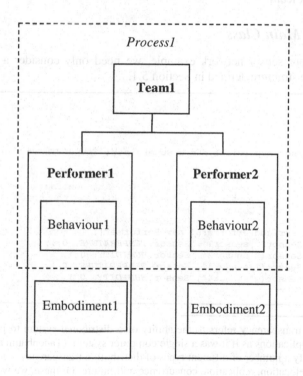

Fig. 5.2 A typical deployment for physically grounded team members

GORITE behaviour models (or process models) as they are more commonly called, can be deployed in a distributed manner. However, behaviours then become more complicated by the need to provide distribution transparency[5]. Our preference is to, where possible, minimise this complexity by consolidation of team behaviours into a single process.

Basic team behaviour was introduced in the examples presented in Sections 5.3 and 5.4. In our sensor network examples, we will explore the issues of task team reformation in response to a dynamically changing environment. We will do this in three stages that expand on the sensor network examples that were employed in Chapter 3:

1. Develop a team-based fault-free sensing application
2. Allow team reformation to occur on sensor failure and
3. Allow team reformation on supervisor failure

In these three applications, the application entry point, team member behaviours and task team deployment are more or less identical, so these are discussed below, separately from the applications. The focus of the individual application discussions then becomes the process models and other functionality required by the supervisor team.

5.5.1 The Main Class

In terms of our sensor network example, we need only consider a single team, which has the structure defined in Section 5.1:

```
// Create a sensor network with 4 sensors and one supervisory
// node.

// Create the supervisory node.
Supervisor supervisor = new Supervisor( "supervisor" );

// Create and add sensors to the network. The sensors are able to
// sense temperature and humidity but are configured to sense
// only one property. The third parameter is the mode:
// 0 = Sensor.Controller.RELIABLE,
// 1 = Sensor.Controller.UNRELIABLE.
supervisor.addPerformers( new Performer[] {
   new Sensor( "sensor1", Sensor.TEMPERATURE, 0 ),
   new Sensor( "sensor2", Sensor.HUMIDITY, 0 ),
   new Sensor( "sensor3", Sensor.TEMPERATURE, 0 ),
   new Sensor( "sensor4", Sensor.HUMIDITY, 0 ),
});
```

---

[5] Distribution transparency refers to the ability of a distributed system to present itself to users and applications as if it was a single computer system. (Tanenbaum and Van Steen, 2007) identify a number of different types of distribution transparency – access, location, migration, relocation, replication, concurrence and failure. Of these, we would argue that failure is the critical transparency from a team perspective – that is, the ability to hide the failure and recovery of a resource. Various technologies are available to assist in achieving this transparency, such as virtual synchrony (Birman, 2005), but these technologies need to be augmented with explicit reasoning in order to provide effective autonomy.

This definition will appear in the application's main() method and is used for all of the sensor network applications in this chapter. The main() method will also need to initiate goal achievement, using the supervisor's sense() method.

5.5.2 The Sensor Class

The team members are of type Sensor, which is defined below. A sensor interacts with its embodiment through a controller object. The Controller class was defined in Chapter 3 and is not reproduced here. The implementation of the Sensor class is straightforward – all that a sensor needs to do is to create a controller for the data source that it has been tasked to sense and to create and add sensing goals for both temperature and humidity to its default capability.

```
public class Sensor extends Performer {
    // sensor sources
    public static final String TEMPERATURE      = "temperature";
    public static final String HUMIDITY         = "humidity";
    // a controller manages embodiment interaction for the sensor
    public Controller c;

    // the Controller class, defined in Chapter 3, goes here
    // timestamp(), defined in Chapter 3, goes here

    // name is the sensor source - "temperature" or "humidity"
    public Goal    read( String name ) {
      return new Goal( "read "+name ) {
        public States execute ( Data d ) {
          if ( !c.read() ) {
            System.err.println(timestamp()+": "+c.name+" failed");
            return States.FAILED;
          }
          System.err.println( c.name+": <"+timestamp()+","+
            c.source+","+c.reading+","+c.count+">");
          return States.PASSED;
        }
      };
    }

    Sensor(String name, String source, int mode) {
      super(name);
      // Create the controller. Name is the sensor name and source
      // is the name of what is to be sensed. Mode is a convenience
      // for introducing faulty behaviour - it can have the values
      // RELIABLE or UNRELIABLE
      c = new Controller( name, source, mode );
      System.err.println( name +" configured for "+source );
      // add goals that a Sensor instance can perform
      // the goal returned by read is named "read "+source
      addGoal( read( source ) );
    }
}
```

5.5.3 The Supervisor Class

In our sensor network examples, variability in behaviour will arise from variability in supervisor behaviour – there will be no change in sensor behaviour or in the initial team structure as defined in the application's main() method. In terms of commonality, in all our examples sensing will be performed by an unregistered task team instance which is assigned to the instance variable providers, which is a member of the Supervisor class:

```
//The task team definition
public TaskTeam providers = new TaskTeam() {
  {
    addRole( new Role( TEMPERATURE_SOURCE, new String [] {
      READ_TEMPERATURE } ) );
    addRole( new Role( HUMIDITY_SOURCE, new String [] {
      READ_HUMIDITY} ) );
  }
};
```

In our examples thus far, we have distinguished between registered task teams (hello world) and unregistered task teams (meter box assembly). However, in both cases the members of the task team remained fixed throughout the application and the default establishment strategy was employed. In this chapter, we will demonstrate how to change task team membership dynamically and also how to customize the task team establishment process.

5.5.4 Version 1: Fault-Free Sensing

We begin our discussion of sensor network teams with the baseline example presented in Chapter 3. Recall that no failures occur in either the sensors or the supervisor. The supervisory node reads the sensors every interval seconds and this continues forever. Reading is modelled as a loop goal which succeeds when its associated end goal succeeds. The termination condition for a loop goal is provided by an end goal – if the end goal succeeds, the loop goal succeeds. In this case, the end goal returns FAILED, as we do not want the loop goal to terminate.

The FAULT_FREE_READING goal that was developed in Chapter 3 is reproduced below:

```
Goal faultFreeReading() {
  return new LoopGoal( FAULT_FREE_READING, new Goal [] {
    new Goal( "read temperature" ) {
      public Goal.States execute( Data d ) {
        // Temperature is constant
        System.err.println( "T = 25" );
        return Goal.States.PASSED;
      }
    },
    new Goal( "read humidity" ) {
      public Goal.States execute( Data d ) {
        // Humidity is constant
        System.err.println( "H = 50" );
        return Goal.States.PASSED;
      }
    },
    new Goal( "pause" ) {
      public Goal.States execute( Data d ) {
        // interval is in seconds - convert it to milliseconds.
        Integer t = (Integer)d.getValue( INTERVAL );
        if ( TimeTrigger.isPending(d,"alarm",t.intValue()*1000) )
          return Goal.States.BLOCKED;
        return Goal.States.PASSED;
      }
    },
    new EndGoal( "reading completed" ) {
      public Goal.States execute( Data d ) {
        // read for ever
        return Goal.States.FAILED;
      }
    }
  });
}
```

All that we need to do to adapt this model to the team version is to replace the reading goals with team goals:

```
Goal basicReading() {
  return
    new LoopGoal( BASIC_READING, new Goal [] {
      new TeamGoal( TEMPERATURE_SOURCE, READ_TEMPERATURE ),
      new TeamGoal( HUMIDITY_SOURCE, READ_HUMIDITY ),
      new Goal( "pause" ) {
        public States execute( Data d ) {
          // interval is in seconds - convert it to milliseconds.
          Integer t = (Integer)d.getValue( INTERVAL );
          if (TimeTrigger.isPending(d,"alarm",t.intValue()*1000))
            return States.BLOCKED;
          return States.PASSED;
        }
      },
      new EndGoal( "reading completed" ) {
        public States execute(Data d) {
          // read for ever
          return States.FAILED;
        }
      }
    } );
}
```

We have also renamed the process model BASIC_READING. Since individual sensor reading is now delegated to the sensors, BASIC_READING can be used as the reading component in all of our subsequent examples.

Achievement of a team goal is delegated to the task team member that has been assigned to fill the role. In the case above, the READ_TEMPERATURE goal is achieved by the team member allocated to the TEMPERATURE_SOURCE role and the READ_HUMIDITY goal is achieved by the team member allocated to the HUMIDITY_SOURCE role. Note that performers of type Sensor have been provided with the ability to achieve these goals.

In this situation, we are assuming that no faults will occur, so the task team (providers) only needs to be established once. This is achieved by invoking the task team's establish() method directly from the supervisor's goal initiation method:

```
public boolean sense( int interval ) {
  // set the sampling interval
  Data data = new Data();
  data.setValue( INTERVAL, new Integer( interval ) );
  // establish the task team
  providers.establish( data );
  // achieve the goal
  return performGoal(
    new BDIGoal( BASIC_READING ), "reading", data );
}
```

The BASIC_READING goal needs to be added to the team's default capability. This is done in the constructor:

```
public Supervisor( String name ) {
  super( name);
  addGoal( basicReading() );
}
```

Sensing can then be initiated with a call like the following in the application's main() method:

```
// Begin sensing with an interval of 5 seconds.
supervisor.sense( 5 );
```

Output like the following will be produced when the application is run with all sensors configured to run in reliable mode:

```
sensor1 configured for temperature
sensor2 configured for humidity
sensor3 configured for temperature
sensor4 configured for humidity
sensor1: <10:54:35,temperature,25,1>
sensor2: <10:54:35,humidity,50,1>
sensor1: <10:54:40,temperature,25,2>
sensor2: <10:54:40,humidity,50,2>
sensor1: <10:54:45,temperature,25,3>
sensor2: <10:54:45,humidity,50,3>
```

Output will be produced indefinitely – only output from the first 15 seconds of execution is shown.

5.5.5 Version 2: Handling Sensor Failure

In this example, we want to demonstrate how task teams can be reformed dynamically. We will do this by allowing a humidity sensor to fail. When the supervisor detects a failure, its strategy to handle the situation will be as follows:

1. Reset the sensor
2. If the fault persists, reform the task team, excluding the failed sensor

The `Controller` class already has the ability built in to cause a sensor to fail. All that we need to do is in `main()`, configure the sensor to operate in unreliable mode (1):

```
new Sensor( "sensor2", Sensor.HUMIDITY, 1 ),
```

If this modification is made to the previous application, sensing fails on the third reading and execution stops:

```
sensor1 configured for temperature
sensor2 configured for humidity.
sensor3 configured for temperature
sensor4 configured for humidity
sensor1: <10:57:37,temperature,25,1>
sensor2: <10:57:37,humidity,50,1>
sensor1: <10:57:42,temperature,25,2>
sensor2: <10:57:42,humidity,50,2>
sensor1: <10:57:47,temperature,25,3>
10:57:47: sensor2 failed
```

The plan now is that when the supervisor detects that the sensor has failed, it will initiate a reset and carry on. We realise this with the ROBUST_READING goal that was developed in Chapter 3:

```
Goal robustReading() {
  return new LoopGoal( ROBUST_READING, new Goal [] {
    new FailGoal( "monitor sensors", new Goal[] {
      basicReading()
    }),
    new Goal( "handle sensor fault" ) {
      public Goal.States execute( Data d ) {
        return Supervisor.this.handleSensorFault( d );
      }
    },
    new EndGoal( "reading completed" ) {
      public Goal.States execute( Data d ) {
        // read for ever
        return Goal.States.FAILED;
      }
    },
  });
}
```

To recap, basicReading(), which will now return the team version of the BASIC_READING goal developed in Version 1, is wrapped in a fail goal. Reading proceeds until a reading goal fails. The fail goal then succeeds and the fault handling goal is pursued. The fault handling goal calls handleFault(), which is now defined as follows:

```
public Goal.States handleSensorFault(Data d) {
  System.err.println(
      "supervisor: handling fault by resetting sensor" );
  Sensor.Controller c =
      (Sensor.Controller) d.getValue( CONTROLLER );
  if ( c.reset() )
    return Goal.States.PASSED;
  System.err.println(
      "supervisor: resetting of sensor unsuccessful" );
  d.replaceValue( FAULT, SENSOR );
  return Goal.States.FAILED;
}
```

Note that the data context has been augmented with elements that will identify the fault type (FAULT) and provide access to the controller for the failed sensor (CONTROLLER).

As with Version 1, task team establishment is performed explicitly in the goal initiation method:

```
public boolean sense( int interval ) {
  // set the data context
  Data data = new Data();
  data.setValue( INTERVAL, new Integer(interval));
  data.setValue( FAULT, NONE );
  data.setValue( CONTROLLER, null );

  // establish the task team
  providers.establish( data );

  // achieve the goal
  return performGoal(
    new BDIGoal(ROBUST_READING), "sensing", data );
}
```

If the application is built at this point, when it is run, the ROBUST_READING goal will fail after the second reset:

```
sensor1 configured for temperature
sensor2 configured for humidity
sensor3 configured for temperature
sensor4 configured for humidity
sensor1: <11:19:48,temperature,25,1>
sensor2: <11:19:48,humidity,50,1>
sensor1: <11:19:49,temperature,25,2>
sensor2: <11:19:49,humidity,50,2>
sensor1: <11:19:50,temperature,25,3>
11:19:50: sensor2 failed
supervisor: handling fault by resetting sensor
sensor1: <11:19:50,temperature,25,4>
sensor2: <11:19:50,humidity,50,1>
sensor1: <11:19:51,temperature,25,5>
sensor2: <11:19:51,humidity,50,2>
sensor1: <11:19:52,temperature,25,6>
11:19:52: sensor2 failed
supervisor: handling fault by resetting sensor
supervisor: resetting of sensor unsuccessful
```

The task team now needs to be reformed when a sensor has failed after the second reset. In this regard, the default establishment strategy – that is the provision of a distinct and singular role filling, with the performer able to achieve all of the goals required by the role regardless of its operational state – is no longer appropriate. Rather, roles need to be allocated to performers that are configured to perform the role and are operational. As observed earlier, the Role.canFill() and Role.canAct() methods are provided for the development of alternative establishment strategies. However, to understand how these methods can be used to achieve our purpose we first need to revisit the process of task team establishment.

Task team establishment involves two phases:

1. A nomination phase, in which team members are nominated for the filling of the roles in the task team. The default nomination strategy (which has been referred to previously as the establishment strategy), is to nominate a **single** candidate for each role. The candidate is chosen on the basis of a sequential scan of team members – the first team member that is able to perform all the goals comprising the role in question is allocated to the role. Furthermore, with the default strategy, a performer can be nominated to fill only one role in the task team. This phase occurs incrementally for registered task teams when performers are added to the enclosing team as nomination is initiated automatically by the `Team.addPerformer()` method. A performer is considered for all registered task teams.

    In contrast, nomination for unregistered task teams is explicit. Typically, it is achieved by invoking the task team's

    ```
    void updateFillers (Vector performers)
    ```

    method. Note that the members of a team are accessible within a team as an instance variable called `performers` which is of type `Vector`. If required, individual team members can be nominated for a task team via the task team's

    ```
    void updateFillers(Performer p)
    ```

    method.

    This phase results in the population of the task team's `fillers` relation, which contains tuples of the form `<role,performer>`. If a task team needs to regenerate its table of potential role fillers for whatever reason, then the entries in `fillers` must first be removed – `clearFilling()` is available for this purpose. `updateFillers()` is then used for the regeneration, regardless of whether the task team is registered or unregistered.

2. A commitment phase, in which nominated candidates commit to a role filling. The default commitment strategy is that all candidates commit unreservedly. It is the combination of the nomination strategy and the commitment strategy that constitutes an establishment strategy. The commitment phase is initiated by the task team's

> **boolean** establish(Data *d*)

method and results in the addition of data elements with the same names as the roles required by the task team being added to the data context. These elements contain the role fillings, which are accessible as capabilities. If a recommitment is required for a task team, then the role filling data elements are automatically cleared prior to the recommitment.

If establish() is able to fill all the roles required by the task team, true is returned. Otherwise, false is returned. Note that if establish() fails, the reason for failure can be determined by inspection of the role filling data elements in the data context.

Establishment behaviour can be customised by overriding the following methods:

Table 5.5 Task Team establishment customisations

Method	Customisation
boolean Role.canFill(Performer p)	Potential suitability of the performer for the role
boolean Role.canAct(Data d,Performer p)	Actual suitability of the performer for the role

In our current situation, what we want to do is to re-establish the task team when a sensor fails. From a modeling perspective, we view this as a recommitment to the current intention by the team members and not as a complete reformation of task team. Thus, we need to change

1. the nomination strategy, so that **all** sensors that are configured to perform a particular sensing role are considered in the commitment phase and
2. the commitment strategy, so that **only** sensors that are operational are considered.

For a performer to be nominated for a role and to be added to the task team's fillers relation, the performer must satisfy the following conditions:

1. it is of type Sensor,
2. it is configured to read the appropriate data source and
3. it is able to perform all the goals required of the role

The first condition is determined using Java's instanceof operator. To determine whether a sensor is configured for a particular data source, recall that each sensor has a controller object (modelled as an instance variable called c) which among other things contains the data source for which the sensor is configured. Note that data sources are designated as "temperature" and "humidity" whereas roles are designated as "temperature source" and "humidity source". Hence some manipulation will be required to compare roles with data sources. Finally, to determine whether or not the performer is able to achieve all the goals required of the role, the performer can employ the following instance method:

boolean hasGoals(String[] *goals*)

This will determine whether or not the specified goals are present in the performer's default capability. As the method will be invoked from within the role instance's canFill() method, it has access to the list of goals that are required for the role. Unsurprisingly, this instance variable is called required.

The above strategy can be realized by overriding the canFill() methods for each of the roles. The code for the TEMPERATURE_SOURCE role is as follows:

```
public boolean canFill(Performer p) {
   if ( !(p instanceof Sensor) )
     return false;
   Sensor s = (Sensor) p;
   // is the performer configured for this role?
   if ( !TEMPERATURE_SOURCE.equals( s.c.source + " source" ) )
     return false;
   // can the performer perform the goals required by the role?
   return p.hasGoals( required );
}
```

As we shall see later, this code, together with similar code for the HUMIDITY_ SOURCE role, will be incorporated into the task team definition.

Next, our desired commitment strategy is that for a sensor to commit to a role, the following three conditions must be satisfied:

1. The sensor is operational
2. The role has not been filled and
3. The sensor is not filling the other role

The first condition is determined by checking the sensor's status – a value of 0 indicates that it is operational. The determination of the remaining two conditions involves accessing the role elements in the data context. In establish(), canAct() is invoked for each suitable candidate and if a candidate is able to commit to the role being filled, the role data element is updated immediately. Consequently, the second condition is determined by checking the role filler data element – if its value is null, then the role has not been filled.

Determination of the final condition involves two tests:

1. if the value of the other role filler data element is null, then the sensor is not filling that role.

2. If the value of the other role filler data element is not null, then the name of the role filler needs to extracted. If this is not the same as the name of the sensor, then the sensor is not filling that role.

The above strategy can be realized by overriding the canAct() methods for each of the roles. The code for the TEMPERATURE_SOURCE role is as follows:

```
public boolean canAct( Data d, Performer p ) {
    // extract role fillers from the data context. If a role
    // filler has not been set, its value will be null.
    Capability c1 = (Capability)
        d.getValue(TEMPERATURE_SOURCE);
    Capability c2 = (Capability)
        d.getValue(HUMIDITY_SOURCE);
    // extract sensor status
    int s = ((Sensor) p).c.status;
    // HUMIDITY_SOURCE is not filled, so no check required for p
    // being the filler
    if ( c2 == null )
        return c1 == null && s == 0;
    // HUMIDITY_SOURCE is filled, so check if p is the filler
    String name1 = p.name;
    String name2 = ((Performer)(c2.getPerformer())).name;;
    return c1 == null && s == 0 && !name1.equals(name2);
}
```

The code for the HUMIDITY_SOURCE role is similar.

The overridden canFill() and canAct() methods for both roles are incorporated into their respective role definitions. These in turn are incorporated into the task team definition:

```
public TaskTeam providers = new TaskTeam() {
    {
        addRole( new Role( TEMPERATURE_SOURCE,
            new String [] { READ_TEMPERATURE } ) {
            // the definition for canFill() goes here
            // the definition for canAct() goes here
        } );
        addRole( new Role( HUMIDITY_SOURCE,
            new String [] { READ_HUMIDITY } ) {
            // the definition for canFill() goes here
            // the definition for canAct() goes here

        } );
    }
};
```

Finally, we wrap providers.establish() in an ESTABLISHMENT goal. If a task team is unable to be formed, the goal fails and the reason for failure is recorded in the FAULT data element of the data context.

```
public Goal establishment() {
   return new Goal( "establishment" ) {
      public States execute(Data d) {
         System.err.println("supervisor: establishing task team");
         boolean success = providers.establish( d );
         // extract role fillers from the data context. If a role
         // filler has not been set, its value will be null.
         Capability c1 = (Capability)d.getValue(TEMPERATURE_SOURCE);
         Capability c2 = (Capability) d.getValue( HUMIDITY_SOURCE );
         System.err.println(" temperature source "+c1 );
         System.err.println(" humidity source "+c2 );
         if (success)
          return States.PASSED;
         // A task team establishment fault has occurred.
         d.replaceValue( FAULT, TASKTEAM_ESTABLISHMENT );
         System.err.println(
          "supervisor: unable to establish a task team");
         return States.FAILED;
      }
   };
}
```

Note that the establishment goal will only execute the commitment phase and not the nomination phase. The reason for this is that the initial nomination is still valid, since all team members have been nominated for all possible roles that they could fill. In this example, the nomination phase will be explicitly invoked from the goal initiation method prior to goal execution via the `TaskTeam.Update Fillers()` method.

We can now provide the supervisor with an ADAPTIVE_SENSING goal[6]:

```
Goal adaptiveSensing() {
   return new LoopGoal( ADAPTIVE_SENSING, new Goal [] {
      establishment(),
      new EndGoal( "sensing completed", new Goal [] {
         robustReading()
      } ),
      new Goal( "reset data" ) {
         public States execute(Data d) {
            // The role fillings in the data context are
            // cleared when the task team is re-established.
            System.err.println( "resetting data" );
            d.replaceValue( FAULT, NONE );
            d.replaceValue( CONTROLLER, null );
            return States.PASSED;
         }
      }
   });
}
```

which is added to the supervisor's default capability in the usual way.

---

[6] The distinction that we make between a READING goal and a SENSING goal is that with a sensing goal, task team formation is included in the process model whereas with a reading goal, task team formation is external to the process model definition. Thus reading goals can be reused without modification.

What will now happen is that if the ROBUST_READING goal succeeds (which it won't, because reading does not terminate), then the end goal succeeds, the loop is broken and the loop goal succeeds. However, if ROBUST_READING fails, the end goal still succeeds, but the loop is not broken. In this case, the reason for the failure is that a sensor has failed and a software reset has not rectified the problem. The task team needs to be re-established, but first the fault-related data elements need to be reset. This is achieved by the "reset data" goal. On the next iteration of the loop, the task team is re-established when the establishment goal is achieved. This time, canAct() will return false when applied to the malfunctioning sensor. The role filler elements in the data context are automatically cleared prior to task team re-establishment.

Finally, the goal initiation method is as follows:

```
public boolean sense( int interval )         {
    // set the data context
    Data data = new Data();
    data.setValue( INTERVAL, new Integer( interval ) );
    data.setValue( FAULT, NONE );
    data.setValue( CONTROLLER, null );
    // nominate candidates for the task team. Performers is an
    // instance variable of the enclosing team
    providers.updateFillers( performers );
    // achieve the goal
    return performGoal(
        new BDIGoal( ADAPTIVE_SENSING ), "sensing", data );
}
```

Representative output when two faulty humidity sensors are used is shown below:

```
sensor1 configured for temperature
sensor2 configured for humidity
sensor3 configured for temperature
sensor4 configured for humidity
supervisor: establishing task team
    temperature source role filler
chapter5.sensors.v2.Sensor: sensor1
    humidity source role filler
chapter5.sensors.v2.Sensor: sensor4
// 2 sets of readings for sensor1 and sensor 4
removed
    sensor1: <14:01:54,temperature,25,3>
    14:01:54: sensor4 failed
supervisor: handling fault by resetting sensor
// 2 sets of readings for sensor1 and sensor 4
removed
    sensor1: <14:01:56,temperature,25,6>
    14:01:56: sensor4 failed
```

```
supervisor: handling fault by resetting sensor
supervisor: resetting of sensor unsuccessful
resetting data
supervisor: establishing task team
   temperature source role filler
chapter5.sensors.v2.Sensor: sensor1
   humidity source role filler
chapter5.sensors.v2.Sensor: sensor2
   // 2 sets of readings for sensor1 and sensor 2 removed
   sensor1: <14:01:58,temperature,25,9>
   14:01:58: sensor2 failed
supervisor: handling fault by resetting sensor
   // 2 sets of readings for sensor1 and sensor 4 removed
   sensor1: <14:02:00,temperature,25,12>
   14:02:00: sensor2 failed
supervisor: handling fault by resetting sensor
supervisor: resetting of sensor unsuccessful
resetting data
supervisor: establishing task team
   temperature source role filler
chapter5.sensors.v2.Sensor: sensor1
   humidity source null
supervisor: unable to establish a task team
```

Note that when the second humidity sensor fails, task team establishment fails as there are no operational sensors left that can fill the HUMIDITY_ SOURCE role. As a consequence, the ESTABLISHMENT goal fails, which then causes the ADAPTIVE_SENSING goal to fail and the application exits.

5.5.6 Version 3: Handling Supervisor Failure

Until now, the supervisor has been pursuing a single intention – that is the sensing of data at regular time intervals. What we now want to do is to allow a supervisor to fail and to have its orphaned sensors managed by another supervisor. That is, we will dynamically add new members both to the team and to the task team. This process will involve two aspects:

1. The failing supervisor will need to recognize that it is about to fail (that is, available power has dropped below a specified level) and then initiate appropriate actions.
2. A potential supervisor will need to monitor a communications link and when a request to adopt an orphaned sensor arrives, initiate appropriate actions. In parallel with this, the supervisor also needs to be pursuing its primary intention, that of sensing.

The first aspect is essentially a process modelling issue and was considered in Chapter 3 from the viewpoint of a supervisor modelled as a single agent. The second aspect extends our consideration of dynamic team behaviour to address the addition of members to a particular team, in contrast to our previous example that focused on removal. Note that these two aspects are inter-related. In practice, there will be two separate teams that will need to interact in order for sensor transfer to occur. For now, we will assume that a reliable communications infrastructure is available to manage that interaction. Provision of such infrastructure within a BDI environment is still very much a research issue and is one that we touch on in our discussion of future work in Chapter 7.

In this section, we consider both of the above aspects. We first revisit power monitoring but from a team perspective and then we consider the handling of orphaned sensors.

Part 1: Power Level Monitoring

The team version of power aware sensing is directly analogous to the single performer version of Chapter 3. The power monitoring branch remains the same, but the sensing branch is now achieved using adaptiveSensing() rather than robustReading():

```
Goal powerAwareSensing() {
   return new ParallelGoal( POWER_AWARE_SENSING, new Goal [] {
     adaptiveSensing(),
     new Goal( "monitor power level" ) {
       public States execute( Data d ) {
         // power is an instance variable of an enclosing class
         // and is therefore in scope
         int t = ((Integer) d.getValue( THRESHOLD )).intValue();
         if ( power > t )
           return States.STOPPED;
         d.replaceValue( FAULT, POWER_LEVEL );
         return States.FAILED;
       }
     }
   });
}
```

Before we can build and run this application, we first need to have power change with time. Then we need to trap failure of the POWER_AWARE_SENSING goal and initiate an appropriate action.

As indicated in the process model above, power is modelled as an instance variable. To model its dynamic variability, we will initiate a separate thread that decrements power every second. This functionality is provided to the supervisor via the following method:

```
// consume power very quickly
void start() {
  power = 100;
  Thread t = new Thread (new Runnable() {
    public void run() {
      try {
        while ( power > 0 ) {
          Thread.sleep( 1000 );
          power -= 10;
          System.err.println(
               "supervisor: power level = "+power );
        }
      }
      catch( Exception e ) {}
    }
  });
  t.start();
}
```

To handle failure, the POWER_AWARE_SENSING goal is wrapped in a fail goal
and a fault handling goal is provided:

```
Goal sensing() {
  return new Plan ( SENSING, new Goal[] {
    new FailGoal ( "trap power failure" , new Goal [] {
      powerAwareSensing(),
    }),
    new Goal( "handle power failure" ) {
      public Goal.States execute( Data d ) {
        if ( POWER_LEVEL.equals( (String) d.getValue(FAULT) ) ){
          System.err.println(
            "supervisor: advising supervisors of loss of power");
          System.err.println(
            "supervisor: notifying base of loss of power");
          return States.FAILED;
        }
        // other terminal faults have already been handled
        return States.FAILED;
      }
    }
  } );
}
```

Prior to goal initiation, data elements need to be set and power consumption
started:

```
public boolean sense( int interval ) {
   // set the data context
   Data data = new Data();
   data.setValue( INTERVAL, new Integer( interval ) );
   data.setValue( FAULT, NONE );
   data.setValue( CONTROLLER, null );
   data.setValue( THRESHOLD, 40 );

   // initiate power consumption
   start();
   // achieve the goal
   return performGoal(
      new BDIGoal( SENSING ), "sensing", data );
}
```

Representative output for the application is shown below:

```
sensor1 configured for temperature
sensor2 configured for humidity
sensor3 configured for temperature
sensor4 configured for humidity
supervisor: establishing task team
   temperature source role filler
chapter5.sensors.v3a.Sensor: sensor1
   humidity source role filler
chapter5.sensors.v3a.Sensor: sensor2
sensor1: <19:41:54,temperature,25,1>
sensor2: <19:41:54,humidity,50,1>
supervisor: power level = 90
supervisor: power level = 80
sensor1: <19:41:59,temperature,25,2>
sensor2: <19:41:59,humidity,50,2>
supervisor: power level = 70
supervisor: power level = 60
supervisor: power level = 50
sensor1: <19:42:04,temperature,25,3>
19:42:04: sensor2 failed
supervisor: handling fault by resetting sensor
sensor1: <19:42:04,temperature,25,4>
sensor2: <19:42:04,humidity,50,1>
supervisor: power level = 40
supervisor: advising other supervisors of loss of
power
supervisor: notifying base station of loss of power
supervisor: power level = 30
supervisor: power level = 20
supervisor: power level = 10
supervisor: power level = 0
```

Part 2: Handling of Orphaned Sensors

We begin by observing that since our focus is on team modelling, exactly how a supervisor receives a request is not important. What is important, however, is that the monitoring of requests occurs concurrently with sensing. As we have already seen, GORITE provides the ability to model concurrent intentions through the use of parallel goals. However, in this instance we would argue that monitoring and sensing are loosely coupled and as such are better modelled as two separate intentions rather than as a single composite intention. Consequently, requests from other supervisors will be modelled as an asynchronous event stream. In GORITE, such events (or percepts as they are called) are bound to goals – when a percept is processed, its corresponding goal is initiated and progressed concurrently with any other active intentions.

Note that we are dealing with two separate intentions, so they have separate data contexts, rather than sharing a single data context. In this example, communication between the two intentions will be through a supervisor instance variable:

```
// A flag to indicate to the main intention that the task team
// candidate set has been modified
boolean providersModified = false;
```

As discussed in Chapter 4, the processing of percepts is mediated by instances of the `Perceptor` class. A perceptor is bound to a performer and to a goal for handling percepts. In this application, the performer is the supervisor team and the perceptor is created as an instance variable of the supervisor team:

```
// A perceptor for requests.
public Perceptor
  requestHandler = new Perceptor( this, HANDLE_REQUEST );
```

The goal that this perceptor will initiate when it mediates percept processing is named HANDLE_REQUEST. A ToDo group can be optionally specified to manage the goal execution. One isn't specified, so the default percept processing ToDo group "percepts" is used.

Percept processing is initiated by the perceptor's `perceive()` method, which in our example, will take two arguments – the event object and a data context. `perceive()` causes the goal bound to the perceptor (in this case HANDLE_REQUEST) to be initiated; the event object is then accessible to the goal's `execute()` method through the data context as a data element named "percept". `requestHandler` initiates percept processing from a separate thread created in the supervisor's `listen()` method, which is presented below. The processing is in response to a single HANDOVER event that is generated after a delay of 2 seconds:

```
void listen() {
  // model hand over requests as an event occurring in a
  // separate thread
  Thread t = new Thread (new Runnable() {
    public void run() {
      try {
        Thread.sleep(2000);
        System.err.println(
          "supervisor: request for handover received");
        requestHandler.perceive(
          new Request(HANDOVER,"sensor9",Sensor.TEMPERATURE,0));
      }
      catch( Exception e ) {}
    }
  });
  t.start();
}
```

Handover events are modelled using a `Request` class:

```
public class Request {
  String type;
  String name;
  String source;
  int mode;

  public Request( String t, String n, String s, int m) {
    type = t;
    name = n;
    source = s;
    mode = m;
  }
}
```

The `HANDLE_REQUEST` intention extracts the request object from the data context and then processes it:

```
addGoal( new Goal( HANDLE_REQUEST ){
  public States execute(Data d) {
    Request r = (Request) d.getValue( "percept" );
    if (r.type.equals( HANDOVER )) {
      System.err.println("supervisor: handling handover request");
      // add the orphan to the team
      Sensor s = new Sensor( r.name, r.source, r.mode );
      s.orphaned = true;
      addPerformer( s );
      // nominate s to fill the role for which it is configured
      providers.updateFillers( s );
        // record the change
      providersModified = true;
      return States.PASSED;
    }
    return States.FAILED;
  }
});
```

In terms of processing, the key steps are that the sensor

1. is added to the team:
   `addPerformer( s ) and`
2. is made available for role filling:
   `providers.updateFillers( s );`

The commitment strategy is the same as before for the sensors that were initially allocated to the supervisor, namely

1. The sensor is operational
2. The role has not been filled and
3. The sensor is not filling the other role

However, a different strategy is employed for orphaned sensors – any orphaned sensor fills the role that it was performing previously. This role filling is in addition to any existing role fillings provided by the native team members. To assist in making this distinction, an additional field (`orphaned`) has been added to the `Sensor` class.

The revised definition of the TEMPERATURE_SOURCE role is as follows:

```
addRole( new Role( TEMPERATURE_SOURCE, new String [] {
  READ_TEMPERATURE } ) {
  int orphans = 0;
  public boolean canAct( Data d, Performer p ) {
    if ( ((Sensor) p).orphaned ) {
      orphans++;
      return true;
    }
    // extract role fillers from the data context. If a role
    // fillerhas not been set, its value will be null.
    Capability c1 = (Capability) d.getValue(TEMPERATURE_SOURCE);
    Capability c2 = (Capability) d.getValue(HUMIDITY_SOURCE);
    // extract sensor status
    int s = ((Sensor) p).c.status;
    // HUMIDITY_SOURCE is not filled, so no check required for p
    // being the filler
    if ( c2 == null )
      return ( c1 == null || orphans > 0 ) && s == 0;
    // HUMIDITY_SOURCE is filled, so check if p is the filler
    String name1 = p.name;
    String name2 = ((Performer)(c2.getPerformer())).name;
    return ( c1 == null || orphans > 0 ) &&
      s == 0 && !name1.equals(name2);
  }
} );
```

The humidity source role definition is modified likewise.

All that remains to be done is to check in the sensing intention as to whether orphans have been added. If they have, then the task team needs to be re-established. We achieve this by adding a "review providers" goal as the first goal of the BASIC_READING loop body:

```
Goal basicReading() {
   return
      new LoopGoal( BASIC_READING, new Goal [] {
         new Goal( "review providers" ) {
            public States execute (Data d) {
               if ( providersModified ) {
                  System.err.println(
                     " supervisor: sensor added - reforming task team" );
                  providers.establish(d);
                  providersModified = false;
               }
               return States.PASSED;
            }
         },
         // . . .
```

The goal initiation method is as before, with role filler nomination for the task team preceding goal execution. When the application is run, output like the following will be produced:

```
sensor1 configured for temperature
sensor2 configured for humidity
sensor3 configured for temperature
sensor4 configured for humidity
supervisor: establishing task team
   temperature source role filler
chapter5.sensors.v3b.Sensor: sensor3
   humidity source role filler
chapter5.sensors.v3b.Sensor: sensor2
// 2 sets of readings for sensor3 and sensor 2 removed
supervisor: power level = 90
sensor3: <17:17:33,temperature,25,3>
17:17:33: sensor2 failed
supervisor: handling fault by resetting sensor
// 2 sets of readings for sensor3 and sensor 2 removed
supervisor: power level = 80
sensor3: <17:17:35,temperature,25,6>
17:17:35: sensor2 failed
supervisor: handling fault by resetting sensor
supervisor: resetting of sensor unsuccessful
resetting data
```

```
   supervisor: establishing task team
      temperature source role filler
chapter5.sensors.v3b.Sensor: sensor3
      humidity source role filler
chapter5.sensors.v3b.Sensor: sensor4
   sensor3: <17:17:35,temperature,25,7>
   sensor4: <17:17:35,humidity,50,1>
   supervisor: request for handover received
   supervisor: handling handover request
   sensor9 configured for temperature
   supervisor: sensors added - reforming task team
   sensor1: <17:17:36,temperature,25,1>
   sensor9: <17:17:36,temperature,25,1>
   sensor3: <17:17:36,temperature,25,8>
   sensor4: <17:17:36,humidity,50,2>
   // remaining output removed
```

In the preceding example, we have used an unregistered task team. If a registered task team is employed, then performers are nominated for role filling when they are added to the team. Thus there is no need to invoke the nomination phase in the goal initiation method through an updateFillers() call. Also, when the orphan sensor is added to the team, it will be automatically nominated to fill whatever roles it is able to perform. So again, there is no need for the updateFillers() call. In terms of the commitment phase, recall that the team's deploy() method returns a goal that invokes establish() for the named task team. This method would be used in place of the establishment() method.

Chapter 6
Belief Management

In the previous chapters, the focus has been on goals – the Desires and Intentions of the BDI model. In particular, we have explored both goal representation and the intention lifecycle and how they are realized in GORITE:

Table 6.1 GORITE support for the intention lifecycle

Aspect	Constructs
Representation	Process models
Selection	Plan context, plan precedence, plan choice goals
Initiation	Percepts, procedural invocation
Progression	ToDo groups, meta-goals

Furthermore, this exploration has taken place within an organizational context where intention realization is a collective effort. Thus a similar exercise has been undertaken for both teams and task teams:

Table 6.2 GORITE support for the team/task team lifecycle

Aspect	Constructs
Representation	Teams, performers, roles, task teams
Formation	Team formation, task team establishment
Initiation	Task team persistence
Progression	Dynamic team/task team reformation/re-establishment

Central to both lifecycles are beliefs. As we have already seen, in GORITE intentions are pursued with respect to a data context which is a collection of named data elements that inform the execution (as in role filler elements) or are transformed by the execution (as in the order objects for our manufacturing examples). In the latter case, the data context supports a business process modelling perspective that is unique to GORITE. In this perspective, all performers involved in a particular goal execution are given access to an evolving

D. Jarvis et al.: Multiagent Systems and Applications, ISRL 46, pp. 125–141.
DOI: 10.1007/978-3-642-33320-0_6   © Springer-Verlag Berlin Heidelberg 2013

body of shared data in much the same way as a workflow results in the progressive transformation of a specific set of business artefacts.

GORITE also supports an agent perspective on goal processing, whereby a performer is able to employ its own beliefs to reason about and guide its behaviour. These beliefs are of more long-term persistence and are supported in GORITE by a tuple-based belief model together with a logic programming infrastructure that provides for query-driven belief reflection. One is of course free to model beliefs using conventional data modeling techniques, but certain elements of the GORITE infrastructure (such as plan context and as we shall see later in this chapter, belief propagation), employ query-driven belief reflection. In addition to belief reflection, GORITE also supports belief propagation. While this is not a direct concern of BDI reasoning, the epistemic reasoning that it supports is an essential component of any practical agent reasoning system, as indicated in Jones and Wray's review of intelligent agent frameworks (Jones and Wray, 2006).

Thus GORITE supports two reasoning models:

1. An intentional reasoning model which addresses both intention selection (the primary focus of the BDI model) and intention progress (using ToDo groups and meta-goals) and
2. An epistemic reasoning model in which beliefs are propagated using inference rules

The two models are coupled. Intentional reasoning may result in the assertion or retraction of beliefs which in turn may trigger belief propagation through the activation of inference rules. Alternatively, epistemic reasoning may trigger intentional reasoning if the beliefs that are modified are subject to internal reflection.

The ontological space that underpins performer reasoning is provided by the performer's capability structure. This space contains the performer's goals, inference rules and beliefs represented as named queries. As we have seen in the earlier chapters, if a performer does not provide an explicit capability structure then its goals and the other elements will be stored in its default capability. Capabilities themselves can be named and can form hierarchies. Thus there is provision within GORITE for performers to operate within a rich ontological space. Both goals and beliefs are accessed by name, so there is also the ability for the capability to provide ontological mapping if required.

In GORITE, beliefs are modelled as named queries, that is, as objects that implement the Inquirable interface[1]. This interface consists of the following methods:

```
Query get( Object[] arguments ) throws Exception
String getName()
```

get() returns a query that can be used to formulate the belief; getName() returns the name of the query.

---

[1] The Relation class introduced in Chapter 2 implements the Inquirable interface.

The addition and retrieval of beliefs to and from a performer's ontological space is achieved via the following methods provided by the `Capability` class:

```
void putInquirable( Inquirable belief )
Inquirable getInquirable( String name )
```

As with the capability's `addGoal()` method, if `putInquirable()` is invoked on a performer instance as opposed to a capability instance, the belief is added to the performer's default capability. Also note that regardless of where in a performer's capability hierarchy `getInquirable()` is invoked,

1. all the performer's named beliefs are accessible
2. there is a single interpretation for each named belief

Beliefs with the same name can exist within different capabilities in a capability hierarchy. When this occurs, different beliefs may be returned depending on where in the hierarchy `getInquirable()` was invoked.

In GORITE, facts are viewed as tuples contained in queriable relations. This modeling stance is supported by the `Relation` class which was introduced in Chapter 2 and is discussed in more detail in the next section. A relation in GORITE thus supports two separate abstractions:

1. Beliefs as queries, which is provided by the `Inquirable` interface discussed above and
2. Facts as tuples, which is provided by the `Store` interface.

The `Store` interface consists of the following methods:

```
boolean add( Object[] values ) throws Exception
boolean remove( Object[] values) throws Exception
void clear() throws Exception
Observable getMonitor()
```

Basic tuple manipulation is provided by the `add()`, `remove()` and `clear()` methods. The interface also provides support for the notification of changes to the store through the `getMonitor()` method.

Note that if a belief has an explicit tuple representation, the convenience method

```
Store getStore( String name )
```

is provided by the `Capability` class to allow access to the belief as a store.

Developers are free to implement alternative belief models to that provided by the `Relation` class. However, while such models must implement the `Inquirable` interface to take full advantage of the GORITE framework, the same is not the case for the `Store` interface. In particular, if a belief can be generalized as a rule, as in

```
low_power( threshold, level )
```

the belief can be defined procedurally within the query itself and there is no need
to reference an associated store.

6.1 Belief Representation

As discussed in the previous section, GORITE supports a belief model that
provides the following abstractions:

1. Beliefs as queries and
2. Facts as tuples

Central to the realization this model is the `Relation` class, which implements
both the `Inquirable` and `Store` interfaces using a logic programming
formulation. Consequently, before we discuss relations, we begin by first
positioning the GORITE belief model within a logic programming context.

6.1.1 Formulation

In GORITE's logic programming formulation of belief, a *relation* is defined as a
set of tuples that have the same attributes. *Facts* are then modelled as tuples and a
predicate is informally defined as a statement about the tuples in a relation that
may be true or false.

Predicates are applied to relations; the relation definition acts as the template
for predicate construction. Thus in Chapter 2, the predicate interpretation of the
relation

```
    can_greet( planet, language, utterance )
```

is that Paul can greet inhabitants of *planet* who speak *language* by uttering
utterance.[2]

The statement constituting a predicate may be posed using either constants,
bound logical variables or unbound logical variables to identify the tuple
attributes. If the statement is posed using only constants or bound variables, we
will refer to it as a *fact* – the term *predicate* is retained for statements that
incorporate unbound logical variables. Such statements can be viewed as
relational queries that return true or false, together with all possible bindings for
the unbound variables through a process of unification.

Queries involving multiple relations can be composed using the standard forms
(*and, or, not* etc.). Note that there are situations in which we are not interested in
the values of particular unmatched attributes. For example, if we wanted to know
whether Paul spoke any earth languages, we don't need to know the values of the

---

[2] One could more properly represent this predicate as the conjunction of two simpler
predicates, `spoken_by( planet, language )` and `can_greet(language,
utterance)` but the given form was used for simplicity of presentation.

language and utterance attributes. In logic programming, the anonymous variable (_) is used in such situations, and the corresponding predicate would be `can_greet( "earth",_,_ )`.

To summarise, the correspondence between logic programming concepts and their GORITE counterparts are as follows:

Table 6.3 Mapping of logic programming concepts to GORITE concepts

Logic Programming Concept	GORITE Concept	GORITE Realisation
Predicate	A query on a relation	`Relation.get()`
Fact	A tuple that has been asserted or retracted on a relation	`Tuple` instance
Logical variables	Logical variables	`Ref` instance
Anonymous variable	Anonymous variable	`null`

6.1.2 Relations

In GORITE, a relation is created by specifying the types of the tuple members or the number of members, in which case the members are of type `Object`:

```
Class c[] = { String.class, String.class, String.class };
can_greet = new Relation( "can greet", c );
```

or

```
can greet = new Relation( "can greet", 3 );
```

A relation can have one or more key fields, in which case the infrastructure will ensure that that only tuples which are unique with respect to their key fields are added to a relation. For the "can greet" relation, the three fields were:

1. planet
2. language
3. greeting

By default, all members of the tuple form the key. The key fields are specified for a relation using the addConstraint() method; fields which are key fields are designated as true; others are designated as false. The general form of this method takes a single argument of type boolean[]:

```
void addConstraint( boolean[] pattern )
```

but simpler forms are available for the cases where the tuples consist of 1, 2 or 3 members:

```
void addConstraint( boolean x )
void addConstraint( boolean x1, boolean x2 )
void addConstraint(boolean x1, boolean x2, boolean x3)
```

Thus for the can_greet relation, the default condition of all tuple members being key fields could be explicitly specified as either

```
can_greet.addConstraint( new boolean[] { true, true, true } );
```

or

```
can_greet.addConstraint( true, true, true );
```

In GORITE, the Relation class provides a

```
Query get(Object[] values)
```

method for querying relations[3]. The arguments for a query correspond to the fields in the relation being queried and are designated as either inputs or outputs. Inputs are either constants or bound Ref objects; outputs are unbound Ref objects. Note that value assignment for a Ref object is managed internally by the cursor object and not by the application.

As noted in Chapter 2, the definition of a query (e.g. its specification using get()) is separate from its binding, which is generated by multiple calls to

```
boolean next()
```

---

[3] What is actually returned is an instance of the Relation.Cursor class, which implements the Query interface.

If a new binding can be generated, `true` is returned; if not, `false` is returned. The bindings themselves are made available in a vector which is supplied to the query via its

```
Vector getRefs(Vector v)
```

method. If a rebinding of the query is required, then its `reset()` method must be invoked prior to any subsequent calls to `next()`.

As our first example, consider the `can_greet` relation of Chapter 2. If it was to be normalized by using the two relations

```
speaks(planet,language)
```

and

```
greets(language,greeting),
```

then a compound query is required to identify the greetings that can be employed for a particular planet. In this case, the individual queries on the two relations need to be joined on the common language attribute to generate a query of the form:

speaks(planet, $x) and greets($x, $y)

This is realized in GORITE using an And query, as shown below:

```
// Create the speaks relation. Tuples are of the form
// <planet,language>. Both members are key fields.
Relation speaks = new Relation( "speaks", 2);
speaks.add( "mars", "martian" );
speaks.add( "earth", "english" );
speaks.add( "earth", "french" );

// Create the speaks relation. Tuples are of the form
// <language,utterance>. Language is the key field.
Relation greets = new Relation( "greets", 2);
greets.addConstraint( true, false );
greets.add( "martian", "VDREW^% ^%FD$^");
greets.add( "english", "hello world");
greets.add( "french", "bonjour le monde");
greets.add( "swedish", "tjena moss");

// Create the query. No binding has occurred.
Ref $language = new Ref( "$language" );
Ref $greeting = new Ref( "$greeting" );
String planet = "earth";

Query q = new And( new Query[] {
  speaks.get(new Object []{planet, $language } ),
  greets.get(new Object []{$language, $greeting}),
});
```

```
Vector<Ref> refs = q.getRefs( new Vector<Ref>() );

// Bind and review the Ref objects.
System.err.println( "Query = " + q );
System.err.println( "Bindings are:" );

int b = 0;

while ( q.next() ) {
  System.err.println(" binding "+b++ );
  for ( Iterator<Ref> i = refs.iterator(); i.hasNext(); ) {
    Ref r = i.next();
    System.err.println( "    Ref " + r.getName() + " = " + r );
  }
}
```

Alternatively, the relations could have beeen added to the performer's default capability as either unpopulated inquirables

```
putInquirable( new Relation( "speaks", 3 ) );
putInquirable( new Relation( "greets", 2 ) );
```

or as populated inquirables:

```
// Create and populate the speaks and greets relations here

putInquirable( speaks );
putInquirable( greets );
```

In the former case, the relations can be populated by retrieving the relations as stores via getStore() and then invoking Store.add().

To obtain queries for the relations, one would first retrieve the relations as inquirables:

```
Inquirable ispeaks = getInquirable( "speaks" );
Inquirable igreets = getInquirable( "greets" );
```

Inquirable.get() can then be used to return a query.

6.2 Belief Reflection

As indicated in Chapter 4, percept processing can be triggered by internal reflection as well as external observation. Internal reflection uses the `Reflector` class to monitor the performer's beliefs and to trigger percept processing on situational change. Monitoring of beliefs by the reflector is achieved through one or more queries, which are added to the reflector using its `addQuery()` method. If a goal is specified in addition to the query in the method invocation, then that goal is pursued when the query changes state, either from false to true or from true to false. If a goal is not specified, then the goal specified when the reflector was constructed is used. The constructor has the following form:

```
Reflector( Performer p, String goal, String todo )
```

Note that unlike perceptor construction, there are no defaults for either the goal name or the ToDo group.

As an example of reflector usage, consider the power monitoring example of Chapter 5 in which power was modelled as a member variable and monitored using a parallel goal as follows:

```
Goal powerAwareSensing() {
  return new ParallelGoal( POWER_AWARE_SENSING, new Goal [] {
    adaptiveSensing(),
    new Goal( "monitor power level" ) {
      public States execute( Data d ) {
        // power is a member variable of an enclosing class
        // and is therefore in scope
        int t = ((Integer) d.getValue(THRESHOLD)).intValue();
        if ( power > t )
          return States.STOPPED;
        d.replaceValue( FAULT, POWER_LEVEL );
        return States.FAILED;
      }
    }
  });
}
```

The updating of `power` was managed in a separate thread.

To use reflection to monitor power usage, we model power as a relation, rather than as an instance variable. The tuples will be of the form *<timestamp, level>*, with timestamp being the key field. The timestamp is a string formed from the following expression:

```
String.format("%tT", System.currentTimeMillis() );
```

The special timestamp "current" is used to designate the current reading. For reasons of simplicity, the current reading is added twice – with timestamps of "current" and of the actual time. As before, updating occurs in a separate thread:

```
void start() {
   // consume power
   Thread t = new Thread (new Runnable() {
      public void run() {
         int p = 100;
         try {
            power.add( timestamp(), p );
            power.add( "current", p );
            while ( p > 0 ) {
               Thread.sleep( 2000 );
               p -= 10;
               power.add( timestamp(), p );
               power.add( "current", p );
               System.err.println( "start: power level = "+p );
            }
         }
         catch( Exception e ) {}
      }
   });
   t.start();
}
```

The power relation is created in the supervisor's constructor as follows:

```
// Beliefs regarding power level
// tuples are of the form <timestamp,level>
@SuppressWarnings( "rawtypes" )
Class c[] = { String.class, Integer.class };
power = new Relation( "power", c );
power.addConstraint(true,false);
```

The reflector is also created in the constructor:

```
r = new Reflector( this, "low power", "reflections" ) {
   {
      final Ref p = new Ref( "power" );
      addQuery(
         new And( new Query [] {
         power.get( new Object[] { "current", p } ),
         new Condition() {
            public boolean condition() {
               int level = ( (Integer) p.get() ).intValue();
               System.err.print(
                  "reflector "+timestamp()+": power level = "+p);
               if (level > THRESHOLD) {
                  System.err.println(" - above threshold");
```

```
            return false;
        }
        System.err.println(" - below threshold");
        return true;
        }
    }
    })
    );
    }
};
```

The reflector has a single query that will trigger the reflector's default goal (low power) when the current power level changes and the level drops below the threshold value. The low power goal is responsible for ensuring that the supervisor shuts down gracefully. The key aspect of this shutdown is the communication with the base station and nearby supervisors:

```
addGoal( new Goal( "low power" ){
  public States execute(Data d) {

    // communication with base station and nearby supervisors
    // goes here

    System.err.println( timestamp()+": shutting down" );
    return States.PASSED;
  }
});
```

In this instance, we choose not to stop the sensing goal, which now is independent of power monitoring, but rather, let it continue until the supervisor has lost all power:

```
Goal adaptiveSensing() {
  return new LoopGoal( ADAPTIVE_SENSING, new Goal [] {
    establish(),
    new EndGoal( "sensing completed", new Goal [] {
      robustReading()
    } ),
    new Goal( "reset data" ) {
      public States execute(Data d) {
        // The role fillings in the data context are
        // automatically cleared when the task team is
        // re-established.
        System.err.println( "resetting data" );
        d.replaceValue( FAULT, NONE );
        d.replaceValue( CONTROLLER, null );
        return States.PASSED;
      }
    }
  });
}
```

6.3   Belief Propagation

In GORITE, beliefs can be propagated using *if/then* rules. The rules are represented in terms of queries – an antecedent query (the *if* part) and a consequent query (the *then* part). Propagation is then achieved through unification of the logical variables that are employed in the two queries. Unification is iterative in the sense that application of a rule to a belief state may result in the assertion of new beliefs or the retraction of existing beliefs. These changes may then trigger another unification cycle with the process continuing until the belief state is unchanged.

As indicated earlier in this chapter, the expectation is that epistemic reasoning using if/then rules will operate in parallel with intentional reasoning. Consequently, belief propagations is performed exhaustively by the executor after each plan step.

As indicated above, rules are represented as queries and are added to a performer's capability using the capability's addRule() method, which has the following form:

void addRule(Query *antecedent*, Query *consequent*)

As with addGoal(), if the rule is not added to a specified capability within the performer, the default capability is used. Note that capabilities can be used to conceptually group related rules (and/or goals).

As an example of belief propagation, consider a system that consists of a number of interconnected components that have well defined inputs and outputs, such as the fuel system for a car. Furthermore, assume that faulty behaviour can be easily assigned to an input/output stream, but not to a particular component. For example, if no petrol is reaching the carburetor, then the fault could lie with the carburetor or with any component of the fuel delivery system between the carburetor and the fuel tank. From a diagnostic viewpoint, what we want to do is to generate a list of components that could be the cause of the fault and will then be the subject of further testing. These candidate components will be the ones that directly or indirectly influence the component whose output was observed to be faulty. This process will be managed by a performer of type Monitor.

We will first model the influence network for the system as the relation

affected_by(component1, component2)

by which we mean that component1 is affected by component2. The first field is the key field.

The relation is populated as follows:

```
affected_by.add( "a", "b" );
affected_by.add( "a", "c" );
affected_by.add( "c", "d" );
affected_by.add( "c", "e" );
affected_by.add( "f", "c" );
affected_by.add( "f", "b" );
```

The candidates will be stored in a separate relation:

```
candidates( component )
```

When a potential fault candidate is added to this relation, a propagation rule will be automatically triggered. This rule will add to the candidates relation all components that could have caused cause the nominated candidate to be faulty. In predicate logic, this rule would have the following form:

and(candidate($x), affected_by($x, $y)) => candidate($y)

In query form, the antecedent is

```
Query q1a = new And(
  new Query [] {
    candidates.get( new Object [] { $x } ) ,
    affected_by.get( new Object [] { $x, $y } ),
    new Condition() {
      public boolean condition() {
        System.err.println( "q1a: <"+$x.get()+","+$y.get()+">");
        return true;
      }
    },
  }
);
```

and the consequent is

```
Query q1b = new And(
  new Query [] {
    // asserts $y when the rule is fired
    candidates.get( new Object [] { $y } ),
    new Condition() {
      public boolean condition() {
        System.err.println( "q1c: <"+$y.get()+">");
        return true;
      }
    },
  }
);
```

$x and $y are logical variables defined as follows:

```
Ref $x = new Ref( "$x" );
Ref $y = new Ref( "$y" );
```

The condition queries above are not essential – they are used solely for tracing purposes. Also note that the statement

```
candidates.get( new Object [] { $y } ),
```

defines the predicate *candidate($y)*, as indicated earlier in Table 6.3. It is only when the rule is fired that bindings are generated. In this case, $y will have the bindings that were generated for the antecedent. As a result, the consequent predicate becomes a fact that is asserted and is added to the candidates relation.

The propagation rule is added to the performer's default capability with the following statement:

```
addRule( q1a, q1b );
```

To exercise the rule, we need a test plan:

```
addGoal(
  new Plan( "test propagation", new Goal [] {
    new Goal("add first candidate") {
      public Goal.States execute( Data d ) {
        Monitor.this.dump( candidates );
        try {
          candidates.add( "f" );
        }
        catch ( Exception e ) {}
        return States.PASSED;
      }
    },
    new Goal("check propagation") {
      public States execute( Data d ) {
        Monitor.this.dump(candidates);
        return States.PASSED;
      }
    },
  })
);
```

which produces the following output when run:

```
candidates:

 qla:  <f,b>
 qla:  <f,c>
 qla:  <f,b>
 qla:  <f,c>
 qla:  <c,d>
 qla:  <c,e>
 qla:  <f,b>
 qla:  <f,c>
 qla:  <c,d>
 qla:  <c,e>

candidates:
 <f>
 <d>
 <e>
 <b>
 <c>
```

Note that the antecedent (q1a) is activated 3 times in the above example:

1. when f is asserted because of the `candidates.add()` statement in the test plan
2. when b and c are asserted as the consequence of the first firing of the rule firing and
3. when d and e are asserted as a consequence of the second firing of the rule

For each rule firing, the matching on the antecedent uses all members of the candidates relation, so previous matchings are regenerated.

In the above example, belief propagation in one relation (`affected_by`) results in the assertion of beliefs in another relation (`candidates`). If appropriate, beliefs can of course be asserted in the same relation.

Propagation can also result in belief retraction through the use of Not queries. A Not query returns true if the query fails and if it appears in a consequent, its bindings are retracted. Beliefs that have been retracted constitute additional information that is maintained by a relation. This information is accessible through a Lost query, which returns true if the query matches beliefs that have been retracted.

As an example of belief retraction, suppose that in the previous example, rather than pruning the search space by identifying a component that could be faulty, we identify links that are not faulty. What we may then want to do is to retract beliefs from the `affected_by` relation that correspond to correctly functioning paths. The predicate form for a rule to achieve this is the following:

and(lost(affected_by($x, $y)), affected_by($y, $z)) => not(affected_by($y, $z))

To illustrate how the rule works, consider the *affected_by* predicate defined previously. Now suppose that the link *<f,c>* was removed because we had determined that it was working correctly. Therefore, the *lost* predicate will generate the binding *<$x = f, $y = c>*. The second part of the antecedent will then generate two bindings, *<$y = c, $z = d>* and *<$y = c, $z = e>* and the nett result of the first firing of the rule antecedent are the two bindings *<$x = f, $y = c, $z = d>* and *<$x = f, $y = c, $z = e>*. By virtue of the *not* predicate, the facts *affected_by(c, d)* and *affected_by(c, e)* are then retracted. This will then cause a second firing of the antecedent, but no new bindings will result.

The corresponding query for the antecedent is

```
Query q2a = new And(
  new Query [] {
    new Lost ( affected_by.get( new Object [] { $x, $y } ) ),
    affected_by.get( new Object [] { $y, $z } ),
    new Condition() {
      public boolean condition() {
        System.err.println(
          "q2a: <"+$x.get()+","+$y.get()+">");
        System.err.println(
          "q2a: <"+$y.get()+","+$z.get()+">");
        return true;
      }
    },
  }
);
```

and for the consequent

```
Query q2b = new And(
  new Query [] {
    new Not( affected_by.get( new Object [] { $y, $z } ) ),
    new Condition() {
      public boolean condition() {
        System.err.println( "q2b: <"+$y.get()+","+$z.get()+">");
        return true;
      }
    },
  }
);
```

If the rule is exercised in the same way as in the previous example, the following
output is produced when the link <f,c> is removed in the first goal of the plan:

```
affected by:
  <f,b>
  <f,c>
  <a,c>
  <a,b>
  <c,d>
  <c,e>

q2a: <f,c>
q2a: <c,d>
q2a: <f,c>
q2a: <c,e>

affected by:
  <f,b>
  <a,c>
  <a,b>
```

Note that as the pruning is initiated by link removal, only the paths emanating
from *<f,c>* are removed – the   path *<f,b>* remains. The example above was
employed purely for pedagogical purposes. A more useful pruning strategy would
have been to introduce a `working(node)` relation that accumulated working
nodes. The rule would then be modified to match on the working nodes in the
antecedent and to assert new working nodes in the consequent. One could then add
a second rule that propagates backwards from the working nodes to identify
further nodes that are working. A complete example that implements this strategy
is available from the GORITE website.

If the rule is executed in the same way as in the previous example, the following output is produced when the link <1,c> is removed in the first goal of the plan:

Note that as the pruning is initiated by link removal, only the paths emanating from c1c2 are removed — the path <A> remains. The example above was employed purely for pedagogical purposes. A more useful pruning strategy would have been to introduce a working rule (node), a solution that accumulated working nodes. The rule would then be modified to match on the working nodes in the antecedent and to assert new working nodes in the consequent. One could then add a second rule that propagates back-tracks from the working nodes to identify further nodes that are working. A complete example that implements this strategy is available from the CORTH website.

Chapter 7
Future Work

The GORITE kernel offers all the core functionality that its predecessors in the BDI lineage have provided, albeit packaged differently. However, the reason for the development of GORITE was not to develop yet another BDI system, but to build a framework that supports the exploration of the team programming paradigm in real world applications **and** is grounded in the BDI model.

This exploration has two interleaved streams – application development and framework development, with application development informing framework development and vice-versa. It is also prefaced on the following assumptions regarding agent-based application development

1. well designed frameworks can significantly impact on the effort required for application development
2. the team programming paradigm can be used to effectively model a wide range of real world applications
3. significant applications can be developed without recourse to specialized development environments and design methodologies

At this early stage we have no reason to seriously question these assumptions. However, these assumptions do provide a structure for the ensuing discussion on future work.

7.1 Framework Design

Lethbridge and Laganière (2005) define a framework as

> "A skeletal software component that performs functions needed by a class of systems, and which is intended to be incorporated into the design of such systems"

This definition makes explicit two fundamental aspects of a framework, namely that it

1. provides generic functionality required by a class of systems/applications and
2. becomes a component of any developed system

D. Jarvis et al.: Multiagent Systems and Applications, ISRL 46, pp. 143–148.
DOI: 10.1007/978-3-642-33320-0_7   © Springer-Verlag Berlin Heidelberg 2013

The class of systems targeted by the GORITE framework are those in which the overall behaviour is best modelled as workflows that involve teams of intelligent agents and where individual behaviour realization (of both agents and teams) conforms to BDI execution semantics. Broadly speaking, the functionality provided by GORITE addresses these two areas of behaviour representation (goal based process models) and behaviour execution (executors ensuring fair and reproducible execution of performers). From a developer's perspective, having a well-defined and comprehensive execution model that encompasses both goal-driven and event-driven execution is a significant value add. Developing application specific, let alone generic agent execution models is a significant undertaking, as multi-agent systems are distributed system and all the issues associated with distributed execution ultimately need to be addressed (Ford et al., 2010).

In GORITE, the details of model execution are encapsulated within the executor class. As can be seen from the examples presented in the earlier chapters, the focus of system development then moves to behaviour representation and the specification of process models. One could argue that traditional BDI frameworks provide similar encapsulation of execution through the BDI execution loop discussed in Chapter 1. However, the BDI execution loop is open with respect to both coordination of behaviour and of beliefs. In contrast, the GORITE executor makes a data context available to all performers participating in a model execution, thereby reducing the need for the developer to explicitly manage intra-performer communication and performer coordination – both significant tasks. Of course, the extent of the benefit that this provides depends on how well the application domain maps to the GORITE execution model.

In terms of the component aspect of frameworks, we note that GORITE is a collection of Java classes that are used to construct Java-based multi-agent applications. There is no separate "agent" language to master both in terms of the functionality that it provides and in how it integrates with a host language. This approach minimizes both the footprint of GORITE and its learning curve. As all GORITE classes are Java classes, there are no surprises – their behaviour is defined by the API. Furthermore, the framework itself is readily extensible through subclassing.

7.2 Team Modelling

7.2.1 Coordinated Behaviour

The team modelling notions of GORITE are syntactically similar to JACK Teams, but there are significant semantic differences. In particular, GORITE provides a richer and more flexible organisational model through its explicit instantiation and persistence of task teams and its attribution of roles and role filling to task teams rather than to teams. This provides an environment in which different approaches to the modeling of team behaviour can be readily explored. Also, one of the interesting nuances of team goals is that a team goal repeats over a role in the same way that a repeat goal repeats over its control variable. This plural role filling means that a team goal is distributed to all role fillers, and they all need to succeed for the team goal to succeed. As an example, the behaviour of the sensors in Version 3 of Chapter 5 – the orphaned sensor is sensed at the same time as the "native" sensor that reads the same source. This behaviour can be customised through the use of control goals, but useage patterns for the modeling of team behaviour in practical situations need to be identified.

The focus in terms of team behaviour of GORITE and before that, JACK Teams, was the orchestration of team member behaviour. The music example at the beginning of Chapter 5 is a good illustration of the overall objective – to coordinate a group of entities to produce a team outcome. In the case of a string quartet, resource contention is generally not an issue – the violonists will not be sharing an instrument. However, in domains such as manufacturing, it becomes a major concern. In the meter box example of Chapter 5, the problem was circumvented by only moving the shared resource (the table) when both assembly streams had requested a move. Such an approach, as we saw in Chapter 5, introduces edge cases that need to be handled. Explicit locking of the table using a semaphore was employed in the original implementation (Jarvis et al., 2006). This results in a "cleaner" solution, but one in which the resource behaviours are tightly coupled because all resource behaviours depend on the semaphore.

Resource contention in the meter box cell is an example of a larger coordination issue within the manufacturing domain, namely how coupled processing streams should be progressed. The reasoning involved, is in general quite complex, as in addition to resource contention, safe operation must be guaranteed and be clearly demonstrable. For example, the table must never move while either a joining, loading or unloading operation is in progress. In (Jarvis et al., 2008b), the concept of next step reasoning was introduced, whereby the cell reviews progress of all interlinked streams on completion of any step. Such an approach is clearly amenable to modeling with ToDo groups – each stream would have a ToDo group that contains the tasks that are currently scheduled for that stream. Effectively, the ToDo group becomes a job list, but for a stream, not an individual machine. The problem is that the stream tasks are specified in terms of roles, not physical machines that can interfere. In the example presented in (Jarvis et al., 2008b), there was a 1-1 mapping between roles and resources, but in general, this will not be the case. How should resource contention be managed in those situations?

The default stance in team programming approaches is that communication between team members is always available and is reliable. We have argued in Chapter 5 that such a stance is reasonable for task team interaction – it makes good design sense to locate the transitive closure of performers and their executors within a single process whenever practical. Inter-process communication can then be handled by separate middleware infrastructure. In the case of management of behaviour/embodiment interactions, one could argue the case for customized solutions, especially if specialized interfacing is required, as is the case with industrial controllers.

Given the above considerations, GORITE provides infrastructure in the form of the RemoteCoaching capability to enable remote services to be modeled as

`RemoteGoal` instances in process models. A remote goal employs the remote service to transform designated inputs into designated outputs. Each service provider will have its own remote coaching capability instance which may contain multiple remote goal instances. When a remote coaching capability instance is constructed, an object of type `RemoteCoaching.Connector` is provided as the argument for the constructor. This object encapsulates the interaction with the remote service provider; remote execution is triggered by invocation of the connector's `performGoal()` method. This method returns an object that implements the `RemoteCoaching.Connection` interface. This interface consists of the two methods, `check()`, which returns the state of the remote execution and `cancel()`, which cancels the remote execution. On the remote side, if the service is provided by a GORITE process model, then the connector/goal interaction would normally be modeled using a perceptor object as described in Chapter 4. From an application perspective, remote interaction is encapsulated within the remote coaching capability and in particular, it is the responsibility of the connection object for a particular goal execution to return the status of the remote goal execution.

If the communications link is unreliable, then that unreliability could be modelled by extending the set of states provided by the `Goal.States` enumeration and returned by the connection object. These values would then need to be assimilated into the process models, either explicitly through additional control overlay goals or implicitly through beliefs. Unreliable communications is the norm in many application domains and its management can become a key driver for architectural design. As an example, UAV missions must be able to cope with loss of communications regardless of whether that loss is unplanned, as in adverse weather conditions or equipment failure, or planned, as in periods of communications silence. As we have noted in Chapter 5, we believe that "reliable" communications is a middleware issue and that there are a range of existing technologies to choose from (Tanenbaum and Van Steen, 2007). The more interesting issue is how a BDI agent functions within such an uncertain communications environment. Critically, it needs to maintain and align models of both the believed state and the actual state of the world. In which direction alignment occurs will depend on the nature of the transactions that have occurred while the agent has been unable to communicate. Clearly some transactions can be rolled back (return to location x) but others can not (missile fired). Also, at some point, the mismatch between the believed state and the actual state will become so great that the current course of action should be terminated and alternatives pursued. At this stage, it is not clear as to how issues such as these should be modelled.

7.2.2 Shared Beliefs

From a research perspective, perhaps the major contribution of GORITE lies in its ability to encapsulate BDI behaviour in the form of executable process models that support a business process modelling perspective. An integral part of this encapsulation is the recognition that a process model operates on a dynamic data context and that the elements of that context are separate from the beliefs that individual performers (and teams) maintain. While belief management at the BDI agent level is well understood, it is not the case with the dynamic data context, which is novel to GORITE. As we have seen in earlier chapters, it provides for considerable flexibility in terms of data modeling, particularly when parallel branches are involved. Further experience with this aspect of modeling is required to better understand what additional support at the framework level is required.

7.2.3 Cognitive Reasoning

As discussed in Chapter 4, one of the weaknesses of the BDI model is that it does not distinguish between tactical reasoning, operational reasoning and strategic reasoning. At a philosophical level, this is not an issue, but at an implementation level, it makes life more difficult than it needs to be. The success of BDI frameworks in addressing real world problems attests to the fact that the BDI model of reasoning is generically applicable. However, from an implementation perspective, frameworks that support richer cognitive models than afforded by the BDI model would make life easier. Note that as demonstrated by Karim and Heinze (2005) in their work on UAV control, the applicability of cognitive reasoning models is not restricted to applications that mimic human behaviour.

GORITE is neutral with respect to cognitive modeling, but what it does offer is a richer representational framework from which to develop such models. A key innovation in GORITE is the ToDo group. From a research perspective, the challenge is then to determine how best to exploit this modeling construct. Should ToDo groups mirror cognitive function, such as the observe, orient, decide and act functions of OODA? Or should they mirror application functionality, such as managing rush orders and executing scheduled orders, as in manufacturing? Or should they do both? Also, how should team behaviour be modelled? Should both teams and team members have separate ToDo groups? How should the interaction be managed?

7.3 Design Methodologies

Design methodologies for agent-based systems are an area of active research but none of the methodologies developed thus far have infiltrated commercial practice. This is unsurprising, given the paucity of commercial agent-based frameworks in the market place. Prometheus (Padgham and Winikoff, 2004) is one methodology which was tied to a commercial BDI system (namely JACK), but it has not seen widespread deployment within the JACK development community. Prometheus does not explicitly address team programming issues, and is therefore unlikely to provide the basis for a suitable design platform for GORITE application development without major redesign.

The underlying premise of Prometheus and other agent-based design methodologies is that the design of agent-based systems is fundamentally different to that of conventional software. If pressed, we might argue that this is not the

case for team programming. Such an argument would centre on the observation
that goal based process models capture the same information as conventional use
cases. This returns us to DCI, which was mentioned in passing in Chapter 1.

The DCI (Data-Context-Interaction) paradigm (Coplien and Bjørnvig, 2010)
was conceived by Trygve Reenskaug, the originator of the MVC (Model View
Controller) pattern. Like team programming, the paradigm is based on the
intuition that complex system behaviour is best specified in terms of
collaborations and roles. In Reenskaug's case this is a long held conviction – he
was an early advocate for role-based modelling, with OOram (Reenskaug, 1995)
and this work had a strong influence on the formulation of collaborations as they
arc presented in UML .

The essence of DCI is that it provides for a separation of what the system is
(the data part, which is in fact the domain model) and what the system does, which
is the interaction part. The interaction part is connected to the data part on an
event-by-event basis through the use of what are called context objects. These
objects encapsulate behaviour modelled as role-based use case scenarios. A
particular context object is responsible for identifying the objects that are required
to play out its scenario and for its activation. Roles are realised as stateless
methods that are defined separately from the class definitions of the objects that
perform the roles. This separation can be achieved using a programming approach
called traits (Schärli et al., 2003); Coplien and Bjørnvig present ways of
mimicking traits for a number of commonly used languages.

The three key differences between DCI and GORITE are that firstly in DCI,
roles are realised as stateless methods, rather than goals, as in GORITE. Secondly,
the objects that participate in a DCI use case scenario are selected by querying
the set of available objects in the domain model. In GORITE, task teams for the
achievement of a goal / performance of a process model are formed from the
available set of performers. However, as task team members are agents rather than
objects, the task team formation process in GORITE allows potential members to
be actively involved in the formation process through negotiation, rather than
being selected, as in DCI. GORITE task teams can also persist beyond a particular
task. Finally, DCI use case scenarios have no direct implementation support – they
are described at the requirements/design levels through use case diagrams and
interaction diagrams, but there is no language-level support for their realisation. In
contrast, GORITE goal decompositions provide an explicit representation of
system behaviour and can be linked to use cases, thus potentially providing a
seamless transition from requirements specification to design and implementation.

In the case of both DCI and GORITE, more experience needs to be gained in
developing role-based applications to determine where the approach fits in the
overall landscape of software development. At this stage, the application of
GORITE is exploratory, identifying applications that might benefit from the
approach and building proof of concept systems. As we progress to the
development of more significant implementations, then customised support in
the form of an IDE would certainly be beneficial. However, how such an IDE
should be structured is another question. In the meantime, we note that GORITE is
a pure Java framework and that it can be used with any Java IDE – the examples
in this book were all developed using Eclipse.

References

AOS Group. AOS™ Autonomous Decision-Making Software (2012),
http://www.aosgrp.com (accessed April 2012)

Birman, K.: Reliable Distributed Systems: Technologies, Web Services, and Applications.
Springer (2005)

Bohemia Interactive, VIRTUAL BATTLESPACE 2 (2012),
http://www.bistudio.com/index.php/english/company/simulations
(accessed April 2012)

Bratman, M.E.: Intention, Plans, and Practical Reason. Harvard University Press (1987)

Bratman, M.: Faces of Intention: Selected Essays on Intention and Agency. Cambridge
University Press (1999)

Brennan, R., Norrie, D.: From FMS to HMS. In: Deen, M. (ed.) Agent Based
Manufacturing. Advances in the Holonic Approach. Springer, Heidelberg (2003)

Brooks, R.: Cambrian Intelligence: The Early History of the New AI. MIT Press (1999)

Bussmann, S., Jennings, N., Wooldridge, M.: Multiagent Systems for Manufacturing
Control. A Design Methodology. Springer (2004)

Cohen, P., Levesque: Teamwork. Nous 25(4), 487–512 (1991)

Connell, R., Lui, F., Jarvis, D., Watson, M.: The Mapping of Courses of Action Derived
from Cognitive Work Analysis to Agent Behaviours. In: Proceedings of Agents at
Work: Deployed Applications of Autonomous Agents and Multi-agent Systems
Workshop, Second International Joint Conference on Autonomous Agents and Multi
Agent Systems, Melbourne, Australia (2003)

Coplien, J., Bjørnvig, G.: Lean Architecture for Agile Software Development. John Wiley
& Sons Ltd. (2010)

Coram, R.: Boyd: The Fighter Pilot Who Changed the Art of War, Little, Brown and
Company (2002)

Decker, K.S.: A Vision for Multi-agent Systems Programming. In: Dastani, M.M., Dix, J.,
El Fallah-Seghrouchni, A. (eds.) PROMAS 2003. LNCS (LNAI), vol. 3067, pp. 1–17.
Springer, Heidelberg (2004)

Deen, M. (ed.): Agent Based Manufacturing. Advances in the Holonic Approach. Springer
(2003)

d'Inverno, M., Kinny, D., Luck, M., Wooldridge, M.: A Formal Specification of dMARS.
In: Rao, A., Singh, M.P., Wooldridge, M.J. (eds.) ATAL 1997. LNCS, vol. 1365,
Springer, Heidelberg (1998)

Evertsz, R., Fletcher, M., Frongillo, R., Jarvis, J., Brusey, J., Dance, S.: Implementing
Industrial Multi-agent Systems Using JACK™. In: Dastani, M.M., Dix, J., El Fallah-
Seghrouchni, A. (eds.) PROMAS 2003. LNCS (LNAI), vol. 3067, pp. 18–48. Springer,
Heidelberg (2004)

Evertsz, R., Ritter, F., Russell, S., Shepherdson, D.: Modelling Rules of Engagement in
Computer Generated Forces. In: Proc. of 16th Annual Conference on Behaviour
Representation in Modelling and Simulation, Simulation Interoperability Standards
Organisation (2007)

Evertsz, R., Pedrotti, M., Busetta, P., Acar, H., Ritter, F.: Populating VBS2 with Realistic Virtual Actors. In: Proc. of 18th Annual Conference on Behaviour Representation in Modelling and Simulation, Simulation Interoperability Standards Organisation (2009)

Flanagan, D.: Java in a Nutshell, 5th edn. O'Reilly (2005)

Ford, K., Allen, J., Suri, H., Hayes, P., Morris, R.: PIM: A Novel Architecture for Coordinating Behavior of Distributed Systems. AI Magazine 31(2), 9–24 (2010)

Fowler, M.: UML Distilled, 3rd edn. Addison-Wesley (2003)

Georgeff, M.P., Lansky, A.L.: Procedural Knowledge. Proceedings of the IEEE 74, 1383–1398 (1986)

Grosz, B., Kraus, S.: Collaborative Plans for Complex Group Action. Artificial Intelligence 86, 269–357 (1996)

Heinze, C., Goss, S., Josefsson, T., Bennett, K., Waugh, S., Lloyd, I., Murray, G., Oldfield, J.: Interchanging Agents and Humans in Military Simulation. AI Magazine 23(2), 37–48 (2002)

Howden, N., Rönnquist, R., Hodgson, A., Lucas, A.: JACK Intelligent Agents – Summary of an Agent Infrastructure. In: Proc. of 5th International Conference on Autonomous Agents, Montreal, Canada (2001)

IEEE Computer Society, The Foundation for Intelligent Physical Agents (2012), http://www.fipa.org (accessed April 2012)

JADE, Java Agent Development Framework (2012), http://jade.tilab.com (accessed April 2012)

Jarvis, B., Jarvis, D., Jain, L.: Teams in Multi-Agent Systems. In: Intelligent Information Processing III. IFIP, vol. 228, pp. 1–10. Springer (2007)

Jarvis, D., Jarvis, J., Rönnquist, R., Howden, N., Newton-Thomas, J.: Smarter Virtual Soldiers for Training and Simulation. In: Proceedings of Land Warfare Conference 2004, Melbourne (2004)

Jarvis, D., Fletcher, M., Rönnquist, R., Howden, N., Lucas, A.: Human Variability in Computer Generated Forces – Application of a Cognitive Architecture for Intelligent Agents. In: Proceedings of SimTectT 2005, Sydney (2005)

Jarvis, J., Jarvis, D., McFarlane, D.: Achieving Holonic Control – an Incremental Approach. Computers in Industry 51, 211–223 (2003)

Jarvis, J., Jarvis, D., Rönnquist, R., Jain, L.: Holonic Execution: A BDI Approach. Springer (2008a)

Jarvis, J., Jarvis, D., Rönnquist, R., Jain, L.: A Flexible Plan Step Execution Model for BDI Agents. Multiagent and Grid Systems 4(4), 359–370 (2008b)

Jarvis, J., Rönnquist, R., McFarlane, D., Jain, L.: A Team-Based Approach to Robotic Assembly Cell Control. Journal of Network and Computer Applications 29, 160–176 (2006)

Jones, R., Laird, J., Nielsen, P., Coulter, K., Kenny, P., Koss, V.: Automated Intelligent Pilots for Combat Flight Simulation. AI Magazine 20(1), 27–41 (1999)

Jones, R., Wray, R.: Comparative Analysis of Frameworks for Knowledge-Intensive Intelligent Agents. AI Magazine 27(2), 57–70 (2006)

Karim, S., Heinze, C.: Experiences with the design and implementation of an agent-based autonomous UAV controller. In: Proceedings of the 4th International Joint Conference on Autonomous Agents and Multi-Agent Systems, Utrecht, The Netherlands (2005)

Kernighan, B., Ritchie, D.: The C Programming Language. Prentice-Hall Inc. (1978)

Koestler, A.: The Ghost in the Machine, Arkana, London (1967)

Laird, J.E., Newell, A., Rosenbloom, P.: Soar: An architecture for General Intelligence. Artificial Intelligence 33(1), 1–64 (1987)

Lethbridge, T., Laganière, R.: Object-Oriented Software Engineering, 2nd edn. McGraw-Hill (2005)

Lui, F., Connell, R., Vaughan, J., Jarvis, D., Jarvis, J.: An Architecture to Support Autonomous Command Agents in OneSAF Testbed Simulations. In: Proceedings of SimTectT 2002, Melbourne (2002)

Millington, I., Funge, J.: Artificial Intelligence for Games, 2nd edn. Morgan Kaufmann (2009)

Moran, P.: OODA Loop, originally drawn by John Boyd (2008), http://en.wikipedia.org/wiki/OODA_loop (accessed April 2012)

Newell, A.: Physical Symbol Systems. Cognitive Science 4, 135–183 (1980)

Newell, A.: The Knowledge Level. Artificial Intelligence 18(1), 87–127 (1982)

OMG, Object Management Group/Business Process Management Initiative. Object Management Group (2012), http://www.bpmn.org (accessed April 2012)

Padgham, L., Winikoff, M.: Developing Intelligent Agent Systems: A Practical Guide. John Wiley and Sons (2004)

Pokhar, A., Braubach, L., Lamersdorf, W.: A Flexible BDI Architecture Supporting Extensibility. In: Proceedings of the 2005 IEEE/WIC/ACM International Conference on Intelligent Agent Technology, IAT 2005 (2005)

Pynadath, D., Tambe, M., Chauvat, N., Cavedon, L.: Toward Team-Oriented Programming. In: Jennings, N.R. (ed.) ATAL 1999. LNCS, vol. 1757, pp. 233–247. Springer, Heidelberg (2000)

Rao, A.S., Georgeff, M.P.: Modelling rational agents within a BDI architecture. In: Allen, J.F., Fikes, R., Sandewall, E. (eds.) Proceedings of the 2nd International Conference on Principles of Knowledge Representation and Reasoning (KR 1991), pp. 473–484. Morgan Kaufman, San Mateo (1991)

Rao, A., Georgeff, M.: BDI Agents: from theory to practice. In: Proceedings of the 1st International Conference on Multi-Agent Systems (ICMAS 1995), San Francisco, California, USA, pp. 312–319 (1995)

Reenskaug, T.: Programming with Roles and Classes: the BabyUML Approach. In: Klein, A. (ed.) Computer Software Engineering Research, pp. 45–88. Nova Publishers (2007)

Reenskaug, T.: Working With Objects: The OOram Software Engineering Method. Prentice-Hall Inc. (1995)

Rönnquist, R.: GORITE, Intendico Pty. Ltd. (2012), http://www.intendico.com/gorite (accessed April 2012)

Rönnquist, R., Jarvis, D.: Interoperability with Goal Oriented Teams (GORITE). In: Fischer, K., Müller, J.P., Odell, J., Berre, A.J. (eds.) ATOP 2008. LNBIP, vol. 25, pp. 118–128. Springer, Heidelberg (2009)

Schärli, N., Ducasse, S., Nierstrasz, O., Black, A.: Traits: Composable Units of Behavior. In: Cardelli, L. (ed.) ECOOP 2003. LNCS, vol. 2743, pp. 248–274. Springer, Heidelberg (2003)

Searle, J.: The Construction of Social Reality, Allen Lane (1995)

Schild, K., Bussmann, S.: Self-Organization in Manufacturing Operations. Communications of the ACM 50(12), 74–79 (2007)

Tamura, S., Seki, T., Hasegawa, T.: HMS Development and Implementation Environments. In: Deen, M. (ed.) Agent Based Manufacturing. Advances in the Holonic Approach. Springer (2003)

Tanenbaum, A., Van Steen, M.: Distributed Systems: Principles and Paradigms, 2nd edn. Prentice-Hall, Inc. (2007)

Thangarajah, J., Padgham, L., Harland, J.: Representation and Reasoning for Goals in BDI Agents. In: Oudshoorn, M.J. (ed.) 25th Australasian Computer Science Conference (ACSC 2002), vol. 4, pp. 259–265. Australian Computer Society, Melbourne (2002)

University of Michigan, Soar (2012), http://sitemaker.umich.edu/soar (accessed April 2012)

U.S. Army, America's Army Official Website (2012), http://www.americasarmy.com (accessed April 2012)

Van Brussel, H., Wyns, J., Valckenaers, P., Bongaerts, L., Peeters, P.: Reference Architecture for Holonic Manufacturing Systems: PROSA. Computers in Industry 37, 255–274 (1998)

Wooldridge, M.: Reasoning About Rational Agents. The MIT Press (2000)

Wooldridge, M.: An Introduction to MultiAgent Systems, 2nd edn. John Wiley & Sons (2009)

Wooldridge, M., Jennings, N.R.: Intelligent Agents: Theory & Practice. The Knowledge Engineering Review 10(2), 115–152 (1995)

Appendix A
The GORITE Tracing Facility

This is a brief discussion about the tracing facility in GORITE, which is enabled when the static member GOAL.debug is set. This may be done in the application code, and also by ensuring the property setting

```
Dgorite.goal.trace=yes
```

before the Goal class is loaded. When tracing is enabled, the GORITE execution thread provides a running commentary, to standard error, about its progress on executing the model. It reports on a number of different things, and in particular how it enters and leaves execution of goal instances.

The execution of a GORITE model may be thought of as a traversal through an abstract tree of pursuits of goals, where each node corresponds to some particular instantiated goal pursuit (or intention), and the sub-node relationship corresponds to the expansion of the goal pursuit into pursuits of sub-goals. Each intention progresses differently depending on the kind of the intended goal, but in general it progresses from a well-defined beginning to an eventual ending, with one or more steps each time the execution thread visits it. In other words, while in execution, an intention moves through a succession of internal states leading from its initial instantiation to its eventual completion as either passed or failed.

The abstract execution thread grows by new sub-nodes being added as the progress of intentions cause sub-intentions. The top level nodes come about for the pursuits of the top level goals. This is typically a BDI goal, which is achieved by means of selecting and performing a suitable plan, and the sub nodes of a BDI goal intention correspond to the attempts of achieving it in the order they are attempted. The execution of a plan is typically like executing a sequence goal, and the sub nodes of an attempted plan instance correspond to the steps of the plan, or more precisely, to the sub intentions that arise for achieving the sub-goals.

The execution trace tells how the execution thread enters and leaves the nodes of the abstract intention tree using path expressions to identify the nodes. Such a path expression starts with the name of the top level goal, and thereafter it identifies the path through sub-nodes by a succession of numbers. The numbers are separated by "*", ":" or "." to indicate whether the sub-node relationship is due to a BDI plan choice, an induced repetition or a sub-goal expansion. The numbering is always from 0 and up, and for example, 4 means the fifth sub-node.

For example, consider the following path expression:

```
[top*1.2:0]
```

This identifies the path from the `top` intention, through the second plan instance, through its third step, to the first of an induced repetition. Thus, the third step of this plan, which defines an attempt to achieve the `top` goal, is a complex goal that involves an induced repetition, such as a loop goal, repeat goal, parallel goal or team goal.

In detail:

- A `SequenceGoal`, and any extension (`EndGoal`, `FailGoal`, `ControlGoal`, `Plan`), results in a path expression affix of ".N", to indicate the sub-goal, counted from 0. Likewise, a `ConditionGoal` yields the same affix to indicate the alternative in enumeration order.
- A `ParallelGoal`, `RepeatGoal`, `LoopGoal` and `TeamGoal` yields an affix of the form ":M.N" where M indicates the (zero based) ordinal for the induced repetition, and N indicates the sub-goal. For a `TeamGoal`, N is always 0, whereas M indicates which filler is concerned.
- A `BDIGoal` is a leaf in a goal hierarchy, but results in an inner node in the abstract execution tree, where it gets associated with the intended plans for achieving the goal. It yields an affix of the form "*N", where N indicates the attempt ordinal (zero based).

When the execution thread enters an intention, the execution trace prints a message line of the form "=> path" to standard error as a note that that intention is entered. When it re-enters an intention, the text "(resume)" is added to the message line, and when it re-enters an intention held in a Todo group, the lead in is by adding a star, making it have the form "=>* path (resume)".

When the exection thread leaves an intention, a message line of the form "<= path (reason)" is printed, where the reason is one of the five reasons why execution abandons the progress of the intention. Namely, that it passed, failed, stopped, blocked or was cancelled.

The execution trace includes additional notes abouet the execution. When the thread enters and leaves a performer are signalled by lead in of ">>" and "<<" respectively, and additional notes are given with lead in of "**" or "--".

As an example, the following is an excerpt of an execution trace from an application (the space travel example provided with the GORITE distribution and not the variant presented earlier in the book). The excerpt includes 20 lines starting at line 81, as follows:

```
81:=>* enterprise:percepts "travel to destination" (resume)
82:** "travel to destination" attempt 1
83:=> enterprise:percepts*0 "travel to destination" (resume)
84:=> enterprise:percepts*0.2 "fly indefinitely" (resume)
85:=> enterprise:percepts*0.2:0 "fly spacecraft" (resume)
86:=> enterprise:percepts*0.2:0:0 "fly spacecraft" (resume)
87:** "fly spacecraft" attempt 1
88:=> enterprise:percepts*0.2:0:0*0 "fly spacecraft" (resume)
89:>> com.intendico.gorite.examples.spacetravel.Martian: jacquie
90:=> jacquie:+percepts*0.2:0:0*0 "fly spacecraft" (resume)
91:Role pilot keeps flying to earth
```

```
92:-- noting jacquie:+percepts*0.2:0:0*0 "fly spacecraft" (BLOCKED)
93:<= jacquie:+percepts*0.2:0:0*0 "fly spacecraft" (BLOCKED)
94:<< com.intendico.gorite.examples.spacetravel.Martian: jacquie
95:-- noting enterprise:percepts*0.2:0:0*0 "fly spacecraft" (BLOCKED)
96:<= enterprise:percepts*0.2:0:0*0 "fly spacecraft" (BLOCKED)
97:-- noting enterprise:percepts*0.2:0:0 "fly spacecraft" (BLOCKED)
98:<= enterprise:percepts*0.2:0:0 "fly spacecraft" (BLOCKED)
99:-- noting enterprise:percepts*0.2:0 "fly spacecraft" (BLOCKED)
100:<= enterprise:percepts*0.2:0 "fly spacecraft" (BLOCKED)
101:=> enterprise:percepts*0.2:1 "stop when arrived" (resume)
```

Line 81	execution resumes the intention named "travel to destination" held at a ToDo group named "percepts" of the performer named "enterprise".
Line 82	is a note to highlight that "travel to destination" is a BDI goal, and at this time its first plan instance is being attempted.
Line 83	execution re-enters the first plan instance of the BDI goal of the "percepts" Todo group of the performer "enterprise", and that it attempts to achieve the goal "travel to destination".
Line 84	execution re-enters the third step of that plan, and that step attempts to achieve "fly indefinitely"
Line 85	execution then enters the first among induced repetitions, which we by knowing the source identify as the first branch of a parallel goal.
Line 86	execution then enters the first among the induced repetitions of the "fly spacecraft" goal. Again by knowing the source, we know this to be a team goal, and therefore this sub-node of the abstract execution tree corresponds to the focussing on the first (and in this case only) role filler.
Line 87	is a note to highlight that "fly spacecraft" is like a BDI goal, and at this time its first plan instance is being attempted.
Line 88	execution then enters that first plan instance. However, since this is due to a team goal, there is an additional transfer goal injected for dealing with the transfer of execution to and from the role filling performer.
Line 89	execution continues into the performer "jacquie".
Line 90	execution enters the first plan instance for achieving the "fly spacecraft" goal, with the subtitle "+" added to the path indicating that it is under a transfer goal.
Line 91	is a trace output from the plan, which is a task plan with a direct implementation in Java.
Line 92	is a note that the plan instance execution method returned in blocked state.

Lines 93 to 100   tells how the execution thread unwinds with the plan instance being
                   blocked up to that the first branch of the parallel goal is blocked.

Line 101   execution then enters the second branch of the parallel goal, and this
                   branch achieves "stop when arrived".

The plans that the tracing involves look according to the following:

For "enterprise":

```
addGoal( new Plan( "travel to destination", new Goal [] {
  new Goal( "percept destination" ) { ... },
    deploy( "travel" ),
    new ParallelGoal( "fly indefinitely", new Goal [] {
      new TeamGoal( "pilot", "fly spacecraft" ),
      new ControlGoal( "stop when arrived", new Goal [] {
          new TeamGoal( "crew", "look out" )
      } )
    } ),
    new TeamGoal( "greeter", "greet" )
} ) );
```

And for "jacquie":

```
addGoal( new Plan("fly" ) {
  public States execute(Data d) {
    System.err.printf(
      "Role %s keeps flying to %s\n",
      d.getValue( BDIGoal.ROLE ),
      d.getValue( DESTINATION ) );
    return BLOCKED;
  }
} );
```

Appendix B
The Key Modelling Classes

Capability

com.intendico.gorite
Class Capability

java.lang.Object
    └─ com.intendico.gorite.Capability

Direct Known Subclasses:
BellBoy, DataForgetting, GoalTransfering, InterIntention, Performer, Performer.RoleFilling, RemoteCoaching, Team.Role, TextQueryCapability

---

public class **Capability**
extends Object

A Capability is a container of Goal hierarchies, which are accessible by their names, and understood to represent alternative ways in which that named goal may be achieved. It provides lookup context for BDIGoal goals.

A Capability may also contain Rule objects, which are added by the addRule(com.intendico.data.Query, com.intendico.data.Query) method. The Rule objects are collated into the owning Performer rule_set to be applied as part of the goal execution.

Nested Class Summary

static interface	**Capability.Deferred** This interface is implemented by the deferred text form entities (rules or reflectors).
class	**Capability.LocalBDIGoal** This is a BDIGoal extension that is tied to the capability such that when executed, only plans of the enclosing capability and its sub capabilities are considered.

Field Summary

Vector	**deferred** The collection of deferred entities.
Hashtable	**goals** The goals of this capability, clustered by name.

Vector	**inner**
	The inner capabilities of this capability.
Hashtable	**inquirables**
	The table of queriable belief structures of this capability, primarily understood qua the `Inquirable` interface.
Performer	**performer**
	The performer that has this capability.
Vector	**rules**
	The `Rule` objects of this capability.

Constructor Summary

Capability()

Method Summary

void	**add**(`Capability.Deferred` entity)
	Utility method to install a Deferred object subsequent to the first invocation of `shareStores(Collection)`
void	**addCapability**(`Capability` c)
	Add an inner capability.
void	**addGoal**(`Goal` g)
	Add a goal to this capability.
void	**addPlan**(`Plan` g)
	Add a plan to this capability.
void	**addReflector**(`String` goal, `String` todo, `Query` query)
	Adds a `Reflector` to this capability.
void	**addRule**(`Query` a, `Query` c)
	Creates a new `Rule` object with given antecedent and consequent `Query` objects.
Goal	**deploy**(`String` name)
	Utility method that creates the goal of establishing a task team.
Inquirable	**getInquirable**(`String` name)
	Returns the performer that has this capability.
Performer	**getPerformer**()
	Returns the Inquirable as mapped, or null if not found.
Store	**getStore**(`String` name)
	Returns an Inquirable qua Store, or null if it's not a Store.
boolean	**hasGoal**(`String` name)
	Utility method to lookup a given goal name.
boolean	**hasGoals**(`String`[] gs)
	Tells whether or not this capability provides some method or methods for all named goals.

void	**initialize**()
	Overridable method where to add Capability initialisation code.
Vector	**lookup**(String name)
	Constructs a vector of all goals under the given name, by considering the local table and recursively from inner capabilities.
void	**putInquirable**(Inquirable inquirable)
	Registers an Inquirable to be mapped.
void	**putInquirable**(String name, Inquirable inquirable)
	Registers an Inquirable under an alias.
void	**setPerformer**(Performer p)
	Sets the performer for this capability, and propagates it to all inner capabilities.
void	**shareInquirables**(Collection shared)
	This method extends the inquirables collection with the given collection, then propagates its map downwards to sub capabilities.

Methods inherited from class java.lang.Object

clone, equals, finalize, getClass, hashCode, notify, notifyAll, toString, wait, wait, wait

Goal

com.intendico.gorite
Class Goal

java.lang.Object
  └ **com.intendico.gorite.Goal**

Direct Known Subclasses:

Action.Usage, Await, BDIGoal, BranchingGoal, ConditionGoal, ContextualGoal, DataGoal, PlanChoiceGoal, RemoteCoaching.RemoteGoal, SequenceGoal, TodoGroupParallel, TodoGroupRoundRobin, TodoGroupSkipBlocked

public class **Goal**
extends Object

The Goal class is base class for a goal oriented process modelling through a hierarchy that reflects the decomposition of a goal as an abstraction of sub goals. A process model is a hierarchy with a top level goal that is sucessively decomposed as particular combinations of sub goals. The top level goal is achieved by means of processing its sub goals according to the goal type.

A goal is processed by instantiating it via its instantiate (java.lang.String, com.intendico.gorite.Data) method, which represents the intention to acieve the goal. The instance gets pursued via its Goal.Instance.perform(com.intendico.gorite.Data) method, which eventually results in invoking the particular Goal.Instance.

`action()` methods of the various extensions, and these implement their particular execution semantics of how their sub goals are processed.

The base class is also used for task goals, where the `execute(com. intendico.gorite.Data, com.intendico.gorite.Goal.Instance)` method is overridden to perform the intended task.

Note that an Goal may be augmented with a reimplemented `instantiate (String,Data)` method as way of defining data values, a reimplemented `execute(Data,Goal.Instance)` method as way of defining post processing, and a reimplemented `cancelled(Goal.Instance)` method as way of restoring state when the goal is cancelled. Using a `BDIGoal` for example, the usage pattern for this may be as follows:

```
new BDIGoal( "the goal to achieve" ) {

    // Define pre-processing at goal instantiation
    public Goal.Instance instantiate(String head,Data data) {
        data.setValue( "name", ..value.. );
        return super.instantiate( head, data );
    }

    // Define post processing after successful instance execution
    public States execute(Data data,Goal.Instance instance) {
        States x = super.execute( data, instance );
        if ( x == PASSED ) {
            // post processing after achieved goal
        }
        return x;
    }

    // Restore if cancelled
    public void cancelled(Goal,Instance instance) {
        // Restore data using instance.setValue(..)
    }
}
```

By that pattern, the data will be set for the `BDIGoal` when it is executed, just prior to creating the execution instance. The `execute(Data,Instance)` method will be invoked one or more times until the goal is achieved (or failed), and the `cancelled(Instance)` will be invoked if the instance execution is cancelled.

Nested Class Summary

class	**Goal.Instance** Base class for goal instance execution procedures.
static class	**Goal.States** These are the return values of the `execute(Data)` method, to report on the state of a goal instance execution.
static interface	**Goal.Tracer** Interface for intrusive tracing of goal execution.

Field Summary

String	**control** The control data, if any.
String	**group** The execution group to be used for pursuing the goal.
String	**name** The name of this goal.
static String	**PERFORMER** The name of the data element that identifies the executing performer.
Goal[]	**subgoals** The sub goals of this goal.
static String	**TRACE** The property name for tracing goal execution.
static Goal.Tracer	**tracer** The global intrusive goal execution tracer, if any.
static boolean	**tracing** This flag captures whether or not the TRACE property set (to any value) upon class initialisation, and it may be set or reset subsequently for controlling the amount of goal execution logging to generate.

Constructor Summary

Goal(String n) Convenience constructor without sub goals.
Goal(String n, Goal[] sg) Constructor.

Method Summary

void	**cancelled**(Goal.Instance which) Control callback whereby the goal gets notified that an (its) intention is cancelled.
Goal.States	**execute**(Data d) This is the simplified execution method for task goal.
Goal.States	**execute**(Data d, Goal.Instance i) This is the primary execution method for a goal.
String	**getType**() Returns a String identifying the type of the goal.
Goal.Instance	**instantiate**(String head, Data d) Utility method for instantiating a goal according to its type.
static boolean	**isTracing**() This is a utility method to discover whether tracing has been requested by defining the TRACE property, or not.

static String	**nameString**(String n) Utility method to quote a string.
String	**toString**() Makes a deep-structure string representation of this goal and all its sub goals.
String	**toString**(String counter) Makes a deep-structure string representation of this goal and all its sub goals, using a given depth prefix.

Methods inherited from class java.lang.Object

clone, equals, finalize, getClass, hashCode, notify, notifyAll, wait, wait, wait

Performer

com.intendico.gorite
Class Performer

java.lang.Object
  └ com.intendico.gorite.Capability
      └ **com.intendico.gorite.Performer**

Direct Known Subclasses:

Team

---

public class **Performer**
extends Capability

The Performer class is base class for Team member implementations. Conceptually a performer is considered to be the agent for Goal executions, and it contains the belief structures, if any, that agent reasoning may refer to.

A Performer is a Capability in that it has Goal hierarchies, and inner capabilities that define its reasoning processes.

The goals of performer may be grouped to belong to named "to-do groups", which are represented by the inner class Performer.TodoGroup. The goals of a TodoGroup are managed such that only one at a time is performed, and the group is associated with meta-level reasoning for deciding which goal to perform next.

Nested Class Summary

class	**Performer.RoleFilling** This class is used for representing this performer when acting in a given Team.Role.
class	**Performer.TodoGroup** The TodoGroup class provides a meta-level goal execution facility, allowing multiple parallel executions be synchronised.

Nested classes/interfaces inherited from class com.intendico.gorite.Capability

Capability.Deferred, Capability.LocalBDIGoal

Field Summary

Executor	**executor** The Executor for this performer.
String	**name** Holds the Performer's name.
Hashtable	**plan_choices** The association between (BDI) goals and their plan choice modifiers.
RuleSet	**rule_set** The Performer's rule set.
static String	**TODOGROUP** The Data.Element name of the Todogroup object concerned when performing TodoGroups meta-level reasoning.
static String	**TODOGROUP_META_GOAL** A constant goal name for a generic TodoGroup meta goal.
Hashtable	**todogroups** This performer's {link TodoGroup} objects.

Fields inherited from class com.intendico.gorite.Capability

deferred, goals, inner, performer, rules, stores

Constructor Summary

Performer() Default constructor.
Performer(String n) Constructor with name given.

Method Summary

void	**addCapability**(Capability c) Overrides Capability.addCapability(com.intendico.gorite.Capability) so as to link up added capabilities with this performer.
void	**addTodoGroup**(Performer.TodoGroup r) Utility method to add a given TodoGroup.
void	**addTodoGroup**(String n, String mg) Utility method to add a TodoGroup with given name and meta goal.
boolean	**changeFocus**() Returns true if the performer execution should change focus.
Goal.States	**execute**(boolean reentry, Goal.Instance instance, String group) Utility method to add an instance to its todogroup.
Performer.RoleFilling	**fillRole**(Team.Role r) Returns a Performer.RoleFilling object for filling a given Team.Role.
String	**getName**() A method that returns name.
Object	**getPlanChoice**(String goal) Utility method to obtain the plan choice modifier, if any, associated with the given goal.
Performer.TodoGroup	**getTodoGroup**(String rn) Utility method to find a named Performer.TodoGroup.
Goal	**goalForPerformer**(Goal g) Utility method to create a TransferGoal object aimed at this performer.
boolean	**performGoal**(Goal goal, String head, Data d) Utility method to instantiate the goal for given the head, and then keep invoking the perform method with the given data context, until it passes or fails.
boolean	**performGoal**(Goal goal, String head, Vector ins, Vector outs) Alternative performGoal method, with input and output held as Vector collections.
boolean	**performGoal**(String goal, String head, Vector ins, Vector outs) Alternative performGoal method, with goal given as a String, implying a BDI goal, and input and output held as Vector collections.
void	**propagateBeliefs**() Utility method to apply all rules exhaustively.
Goal.States	**runPerformer**() Control method to run all todogroups until stopped or blocked.

void	**setPlanChoice**(String goal, Object choice_method) Utility method to declare a plan choice modifier for a goal.
void	**signalExecutor**() Utility method to signal the executor that this performer wants some attention.
static Team	**team**(String rolename, Data d) Utility method to obtain the Team that has established a Team.Role filler under a given name in the Data.
String	**toString**() Returns a text representation of this performer.

Methods inherited from class com.intendico.gorite.Capability

add, addGoal, addReflector, addRule, deploy, getInquirable, getPerformer, getStore, hasGoal, hasGoals, initialize, lookup, putInquirable, setPerformer, shareInquirables

Methods inherited from class java.lang.Object

clone, equals, finalize, getClass, hashCode, notify, notifyAll, wait, wait, wait

Team

com.intendico.gorite
Class Team

```
java.lang.Object
    └ com.intendico.gorite.Capability
        └ com.intendico.gorite.Performer
            └ com.intendico.gorite.Team
```

public class **Team**
extends Performer

The Team class is a base class for a structured task performer, i.e. a Performer that combine sub performers. A particular team type is defined by extending the Team class. The structure of an instance of a team type is defined by calls to the

addPerformer(com.intendico.gorite.Performer)

method. This provides a lookup context for inner Team.TaskTeam objects, which are used to define particular sub team constellations for particular team tasks.

Nested Class Summary

class	**Team.Role** The Role class is a base class for representing team members in a team.
class	**Team.TaskTeam** A TaskTeam represents a sub-grouping of a team's roles as appropriate for a particular team activity.

Nested classes/interfaces inherited from class com.intendico.gorite.Performer

Performer.RoleFilling, Performer.TodoGroup

Field Summary

Hashtable	**groupings** This is a table of TaskTeam objects, to support longer term constallations.
Vector	**performers** This is the current role filling for this team -- it's obligation structure.

Fields inherited from class com.intendico.gorite.Performer

executor, name, plan_choices, rule_set, TODOGROUP, TODOGROUP_META_GOAL, todogroups

Fields inherited from class com.intendico.gorite.Capability

deferred, goals, inner, inquirables, performer, rules

Constructor Summary

Team() Default constructor.
Team(java.lang.String name) Constructor for team of given name.

Method Summary

void	**addPerformer**(Performer performer) Adds a particular performer in a named role filling.
void	**addPerformers**(Performer[] p) Adds multiple, named performers.
Performer	**getPerformer**(int i) Utility method to pick a performer.

Performer	**getPerformer**(java.lang.String name) Utility method to pick a performer by name.
Team.TaskTeam	**getTaskTeam**(java.lang.String name) Utility method to access a long term TaskTeam by name.
Performer	**removePerformer**(java.lang.String name) Utility method to remove a performer by name.
void	**setTaskTeam**(java.lang.String name, Team.TaskTeam group) Utility method to define a long term TaskTeam by unique name.

Methods inherited from class com.intendico.gorite.Performer

addCapability, addTodoGroup, addTodoGroup, changeFocus, execute, fillRole, getName, getTodoGroup, goalForPerformer, performGoal, performGoal, performGoal, propagateBeliefs, runPerformer, signalExecutor, team, toString

Methods inherited from class com.intendico.gorite.Capability

addGoal, addRule, deploy, getPerformer, hasGoal, hasGoals, initialize, lookup, setPerformer

Methods inherited from class java.lang.Object

clone, equals, finalize, getClass, hashCode, notify, notifyAll, wait, wait, wait

Role

com.intendico.gorite
Class Team.Role

```
java.lang.Object
    └ com.intendico.gorite.Capability
        └ com.intendico.gorite.Team.Role
```
Enclosing class:
   Team

---

public class **Team.Role**
extends Capability

The Role class is a base class for representing team members in a team. Role extends Capability, so as to hold Goal methods that are tied to the role within the context of a Team. It further contains an attribute required, which indicate goals that are required by a Performer of the role.

Nested Class Summary

Nested classes/interfaces inherited from class com.intendico.gorite.<u>Capability</u>

`Capability.Deferred`, `Capability.LocalBDIGoal`

Field Summary

`Team.TaskTeam`	**<u>group</u>**	
	The `Team.TaskTeam` that this role is a role of.	
`String`	**<u>name</u>**	
	The reference name for this role.	
`String[]`	**<u>required</u>**	
	The goal names required by a role filler.	

Fields inherited from class com.intendico.gorite.<u>Capability</u>

`deferred`, `goals`, `inner`, `inquirables`, `performer`, `rules`

Constructor Summary

Team.Role(`String` n, `String[]` r)
   Constructor that takes an array of required goals.

Method Summary

`void`	**<u>addCapability</u>**(`Capability` c)	
	Overrides `Capability.addCapability(com.intendico.gorite.Capability)` so as to link up added capabilities with this team qua performer.	
`boolean`	**<u>canAct</u>**(`Data` d, `Performer` p)	
	Overridable method that determines whether the given `Performer` can act in this role within the `intention` represented by the given `Data`.	
`boolean`	**<u>canFill</u>**(`Performer` p)	
	Overridable method that determines if the given performer can fill this role (within its `Team.TaskTeam` group).	
`Team`	**<u>team</u>**()	
	Returns the team of the role.	

Methods inherited from class com.intendico.gorite.<u>Capability</u>

`add`, `addGoal`, `addPlan`, `addReflector`, `addRule`, `deploy`, `getInquirable`, `getPerformer`, `getStore`, `hasGoal`, `hasGoals`, `initialize`, `lookup`, `putInquirable`, `putInquirable`, `setPerformer`, `shareInquirables`

Methods inherited from class java.lang.**Object**
`clone`, `equals`, `finalize`, `getClass`, `hashCode`, `notify`, `notifyAll`, `toString`, `wait`, `wait`, `wait`

TaskTeam

com.intendico.gorite
Class Team.TaskTeam

`java.lang.Object`
   └ `com.intendico.gorite.Team.TaskTeam`

Enclosing class:
   Team

`public class` **Team.TaskTeam**
`extends` `Object`

A TaskTeam represents a sub-grouping of a team's performers into roles as appropriate for particular team activities. Each TaskTeam is populated to hold a selection of performers that are nominated for the task team roles according to their abilities. This role filling defines the candidates for acting in the roles when the TaskTeam is "established" in a particular intention.

The role filling in a TaskTeam as well as the established role filling in an intention may be changed at any time. When the TaskTeam is installed for a `Team`, the current `Team.performers` collection is revide and performers are nominated to roles as appropriate. Thereafter each subsequently added `Performer` is automatically considered, and nominated to roles of installed TaskTeams as appropriate. The detail nomination decision is made by the overridable `Team.Role.canFill(Performer)` method, which by default ensures distinct, singular role filling, and that the nominated `Performer` has plans for all required goals.

The TaskTeam is deployed on demand for intentions, typically by by means of `deploy(String)` goal in team plans. When deployed in this way, the TaskTeam role filling is established in the intention `Data` via `establish(Data)`, with a further filtering through the overridable `Team.Role.canAct(Data,Performer)` method, which by default is "true".

Field Summary

Relation	**fillers**
	The population of this task team, held as a `Relation` between roles and performers.

Vector	**roles** The roles of the task team.

Constructor Summary

Team.TaskTeam()

Method Summary

void	**addRole**(Team.Role r) Utility method to add a role to the task team.
void	**clearFilling**() Utility method to clear all filling.
Goal	**deploy**(String name) Utility method that creates a Goal of establishing this task team in the intention data when the goal is to be achieved.
boolean	**establish**(Data d) Utility method to establish the role filling in the given intention data.
Query	**filling**(Object r, Object p) Utility method to return a role filling Query object for a given role and performer, either of which is given as a constant of its type, null, or a Ref object, which may be (appropriately) bound or unbound.
void	**fillRole**(Team.Role r, Performer p) Utility method to define a role filler for a role.
Hashtable	**getActors**(Data d) Utility method to collate all actors of this TaskTeam's roles in the given Data.
void	**removeRole**(Team.Role r) Utility method to remove a role from the task team.
Team	**team**() Utility method to return the enclosing team wherein this TaskTeam object is created.
void	**updateFillers**(Performer p) Updates the role nominations for the given Performer as appropriate for roles in this taks team.
void	**updateFillers**(Vector performers) Updates the role nominations for the given performers.

Methods inherited from class java.lang.Object

clone, equals, finalize, getClass, hashCode, notify, notifyAll, toString, wait, wait, wait